Elisson Furlan Figueiredo

Molecular epidemiology and genetic diversity of noroviruses

Elisson Furlan Figueiredo

Molecular epidemiology and genetic diversity of noroviruses

associated with viral diarrhea in the Triângulo Mineiro region, Brazil

Imprint

Any brand names and product names mentioned in this book are subject to trademark, brand or patent protection and are trademarks or registered trademarks of their respective holders. The use of brand names, product names, common names, trade names, product descriptions etc. even without a particular marking in this work is in no way to be construed to mean that such names may be regarded as unrestricted in respect of trademark and brand protection legislation and could thus be used by anyone.

Cover image: www.ingimage.com

This book is a translation from the original published under ISBN 978-613-9-62372-3.

Publisher:
Sciencia Scripts
is a trademark of
Dodo Books Indian Ocean Ltd. and OmniScriptum S.R.L publishing group

120 High Road, East Finchley, London, N2 9ED, United Kingdom
Str. Armeneasca 28/1, office 1, Chisinau MD-2012, Republic of Moldova, Europe
Printed at: see last page
ISBN: 978-620-7-72428-4

SUMMARY

Acute childhood gastroenteritis affects more than 700 million children under the age of five every year and causes approximately 1.87 million deaths, a figure that represents 12% of child deaths worldwide. Viral pathogens are among the most common causes of childhood gastroenteritis and norovirus is one of the major causes of sporadic outbreaks of diarrhea in adults and children under five years of age. The aim of this study was to detect and characterize the genetic variability of norovirus in faecal samples from cases of childhood gastroenteritis in the Triângulo Mineiro region, Minas Gerais, Brazil, between 2006 and 2011. Screening was carried out using an enzyme-linked immunosorbent assay and the positive samples were subjected to genotyping by RT-PCR, partial genome sequencing and phylogenetic analysis. A total of 975 fecal samples were analyzed, resulting in 144 (14.7%) positive samples. The prevalence of norovirus was higher than the prevalence of rotavirus (12%), previously characterized in these samples. There was no seasonal pattern related to norovirus infection in the population studied; the most affected age group was 7 to 12 months, with a higher prevalence among cases requiring hospitalization. All the norovirus samples were identified as being from the GII genogroup. Of the 33 noroviruses sequenced, 31 (94%) belonged to the GII.4 genotype, 1 (3%) to the GII.3 genotype, and 1 (3%) to the GII.6 genotype. From the phylogenetic analysis, it was possible to observe that the noroviruses circulating in the Triângulo Mineiro region are very diverse and represent all the GII.4 subgenotypes subgenotypes described in the literature to date, with the greatest number being similar to the 2006b/Nijmegen prototype, followed by the 2004/Hunter subgenotype.

Key words: Gastroenteritis, Acute infantile diarrhea, Norovirus

1   INTRODUCTION

Acute gastroenteritis in children is one of the biggest public health problems worldwide, both in developed and developing countries (BOSCHI- PINTO *et al.*, 2008; LEE et *al.*, 2010; ALAM et al., 2003; TRAN et *al.*, 2010; KHAMRIN *et al.*, 2011). It is a disease characterized predominantly by an increase in the number of bowel movements, with watery stools or stools of little consistency, which can be accompanied by nausea, vomiting, fever and abdominal pain (MINISTÉRIO DA SAÙDE, 2009); as a consequence, it can cause severe dehydration, with loss of fluids and electrolytes which, if not replaced, can, together with other factors, lead to the patient's death. This syndrome is among the main causes of infant morbidity and mortality (UNICEF, 2009) and mainly affects children under five years of age, both in developed and developing countries, where it has a greater impact (GLASS *et al.*, 2000). It is estimated that more than 1 billion cases of acute gastroenteritis occur every year, 700 million of which in children under five years of age alone, leading to more than 1.87 million deaths per year at this age, a figure that represents 12% of child deaths worldwide (BOSCHI-PINTO *et al.*, 2008; UNICEF, 2012). According to data from the World Health Organization in 2009, each child up to the age of five suffers an average of 3.2 episodes of diarrhea per year, but in some countries, this number can reach 10 episodes (WHO, 2009). Diarrhea is a health indicator capable of expressing social inequality, since the disease is more severe in children exposed to poor sanitation conditions, poor nutrition and lack of access to basic health services (LOOPMAN *et al.*, 2004; CHENG *et al.*, 2005). In Brazil, according to data from the Ministry of Health, from 2000 to 2011, 33,397,413 cases of acute gastroenteritis were reported; according to data from the mortality information system, from 2000 to 2009, Brazil had 49,603 deaths from diarrhea associated with gastroenteritis of presumed infectious origin (MINISTÉRIO DA SAÙDE, 2012).

The etiology of gastroenteritis is quite diverse, including bacteria, parasites and viruses (FERNANDEZ AND GÓMES, 2010). Viral pathogens are among the most common causes of gastroenteritis; different types of virus, such as rotavirus (RV), norovirus (NoV), adenovirus (AdEV) and astrovirus (AsTV), among others, are involved in the etiology of non-bacterial diarrheal diseases diagnosed worldwide (CLARK and McKENDRICK, 2004). Although NoVs were the first agents associated with gastrointestinal disease, they have long been considered secondary agents of gastroenteritis, with RVs being considered the most important agents (MORILLO et al., 2011). NoVs, when considering people of all age groups, are the major cause of sporadic outbreaks of non-bacterial diarrhea worldwide (JIANG et *al.*, 1992; *HANSMAN et al.*, 2004; KOOPMANS *et al., 2004;* VASICKOVA *et al.*, 2005; ZINTZ *et al., 2005;* GEORGIADIS *et al.*, 2010; KITTIGUL *et al., 2010*; FIORETTI *et al.*, 2011).

1.1 History

For a long time, researchers tried to elucidate the etiological agent responsible for outbreaks of non-bacterial acute gastroenteritis that caused diarrhea, vomiting and abdominal pain (GREEN *et al.*, 2001). Between 1945 and 1953, the first studies of new etiological agents responsible for outbreaks of acute gastroenteritis were based on inducing the disease in volunteers using faecal lavage filtrates called "bacteria-free faecal lavage" (REIMANN et *al.*, 1945 ; GORDON et *al.*, 1947 ; JORDAN et *al.*, 1953).

The acute diarrheal disease induced after oral administration of a pool of bacteria-free fecal suspension from two patients identified in an outbreak of gastroenteritis at *Marcy State Hospital* in New York, USA, highlighted the viral etiology of this infection (GORDON et *al.*, 1955; PARASHAR et *al.*, 1998); however, all attempts to identify the etiological agent were unsuccessful (WYATT et *al.*, 1974). Therefore, samples collected from sick people continued to be used in experimental studies (DOLIN *et al.*, 1971 ; GREEN *et al.*, 2001). In 1970, a second generation of studies with volunteers began in the USA and the UK with the aim of identifying the etiological agent in these fecal suspensions. Fecal filtrates from four outbreaks of gastroenteritis were studied in volunteers; one of these outbreaks occurred in a high school in Norwalk, Ohio, in October 1968 when, over a two-day period, 50% of the students and teachers developed gastrointestinal illness. The average incubation period and duration of the illness were 48 and 24 hours, respectively. The disease was first described as "Winter Vomiting Disease", and the main clinical manifestations in this outbreak were vomiting and nausea, although some patients developed diarrhea. Laboratory studies have not been able to identify the etiological agent (CUBITT *et al.*, 1975; 1979; ZICKL, ADLER, 1969; GREEN *et al.*, 1995; KAPIKIAN et *al.*, 2000). Bacteria-free fecal suspension filtrates from secondary cases of the Norwalk outbreak induced similar illness in human volunteers. Studies suggested that the causative agent of the disease was a small virus (<36 nm), resistant to ether and stable at high temperatures.

However, it was not possible to propagate this agent in culture or reproduce the disease in an animal model. Subsequently, KAPIKIAN *et* al. in 1972, using immuno-electron microscopy (IME), identified 27 nm viral particles (figure 1), which were characterized as responsible for outbreaks of acute gastroenteritis (CUBITT *et al.*, 1999; KAPIKIAN *et al.*, 1972). These particles were named Norwalk virus, as they resulted from the outbreak of gastroenteritis that occurred in the city of Norwalk, Ohio in 1968. However, the limited availability of viruses in stool samples from patients affected by the disease, and the lack of a cell culture system, limited studies of this agent, and until the early 1980s, the relationship between the epidemiology and classification of these viruses remained uncertain (GREEN *et al.*, 2007).

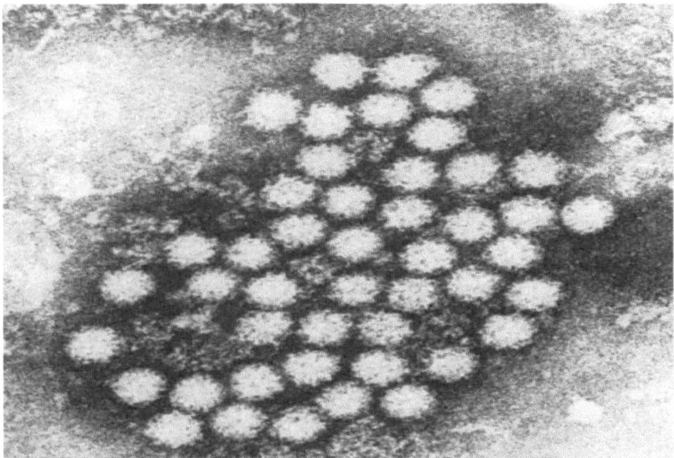

Figure 1: Viral particles aggregated by incubation with human serum in a 1:5 dilution from a patient affected by norovirus (x 31,500). Source: KAPIKIAN et al. (1972).

1.2 Classification and Structure

The Norwalk virus was initially described as belonging to the *Picornaviridae* or *Parvoviridae* families, based on the appearance of the virion using electron microscopy (EM) techniques (CAUL *et al.*, 1996; KAPIKIAN *et al.*, 1996); however, the detection of a structural protein in viral particles purified from human feces was described as a characteristic of the *Caliciviridae family*. The *Caliciviridae* family was distinguished from the *Picornaviridae family* in 1978 (BERKE, 1997) and the name *Caliciviridae* family for caliciviruses was first suggested by the Third International Committee on Virus Taxonomy in 1979. In 1990, the cloning of the Norwalk virus genome confirmed its classification as a member of this family (HARDY & ESTES, 1996). The name *"Caliciviridae"* derives from the Latin word *"calix"* which means *calyx* (GREEN *et al.*, 2000). The family currently consists of four genera: Norovirus and Sapovirus, which represent the human caliciviruses (HuCV), vesiviruses and lagoviruses that infect animals (GREEN *et al.*, 2007).

Initially, the classification of NoV was based on the antigen-antibody reaction using sera from volunteers (GREEN *et al.*, 2007). However, cross-reactivity between samples has shown great antigenic diversity in these viruses. Direct genotyping by neutralization is not possible due to the lack of a viral isolation system in cell culture. Thus, characterization by molecular methods, including reverse transcription followed by polymerase chain reaction (RT-PCR) and especially nucleotide sequencing have been established as the most appropriate tools for characterizing these viruses (ANDO *et al.*, 2000; KATAYAMA et al., 2002; ZHENG *et al.*, 2006).

Morphological analysis of NoV is limited by the low number of viral particles present in the samples. By negative contrast ME, the viral particles show an uncharacteristic outer edge, differentiating NoVs from other caliciviruses (ATMAR & ESTES, 2001). The expression of the NoV capsid protein has been studied in recombinant Baculovirus systems, resulting in the production of *"virus-like particles"* (VLP), which are used for morphological and biochemical studies (BERTOLOTTI-CIARLET *et al.*, 2002; RICHARDS et al., 2003; CHEN *et al.*, 2006).

NoVs have a small, spherical structure, with a diameter of 27 to 40 nm, non-enveloped and with icosahedral symmetry (TAN AND JIANG, 2005; GRENN *et al.*, 2007). Its capsid is made up of 180 protein molecules organized into 90 dimers with a molecular weight of 58kDa (PRASAD, HARD, JIANG, ESTES, 1996), which are made up of two main domains, the S domain which forms the inner part of the capsid that surrounds the genome, the P domain, subdivided into P1 and P2 which form an architecture characterized by its "spiral" shape when observed by ME, also playing a role in the stability of the capsid, and a third domain, N (N-terminal), the innermost (figure 2) (PRASSAD et al., 1999; GREEN et al., 1997, 2000 ; BERTOLOTTI-CIARLET et al., 2002).

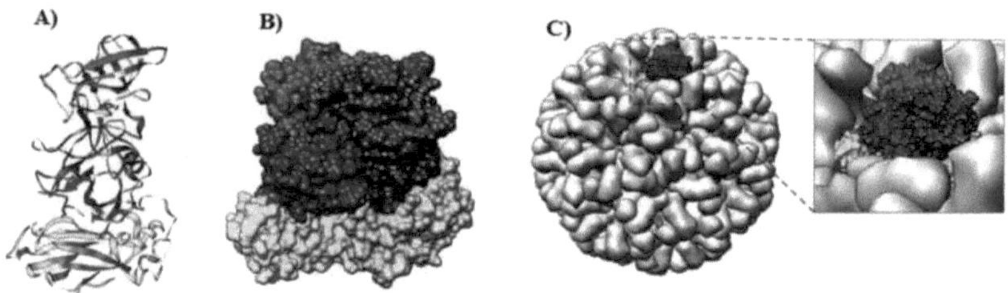

Figura 2: Representation of the monomeric (A) and dimeric (B) structure of the VP1 protein encoded by ORF-2. C) Viral particle showing the location and position of the protein in the particle, formed by 180 monomers of the VP1 protein, which is formed through the linkage between 90 dimers. **Source:** DONALDSON *et al.*, (2010)

The genome of NoV is a single strand of linear RNA with positive polarity and approximately 7.7Kb in length (HARDY & ESTES, 1996). The genome begins with the VPg protein, common to all caliciviruses, covalently linked to the 5'-terminal region of the viral genome, and has been reported to be essential for the infectivity of the virus, playing a fundamental role in initiating the translation of viral RNA (GREEN *et al.*, 2007; DONALDSON *et al.*, 2008). VPg is followed by three open reading regions (ORF-1 , ORF-2 and ORF-3), two untranslated regions (URT) located at both the 5' and 3' ends, and a poly(A) tail in the 3' terminal region (figure 3) (LE-PENDU *et al.*, 2006; *HARDY et al.*, 2005).

At the 5' end is ORF-1, which encodes a polyprotein ranging from 1680 to 1881 aa, which is cleaved into non-structural proteins (KOJIMA *et al.*, 2002; YAN et *al.*, 2003; GEORGIADIS *et al.*, 2010; BORGES *et al.*, 2006). The polyprotein expressed by ORF-1 is cleaved by the 3c-like proteinase (3CLpro), generating at least six proteins: (a) p48, involved in the viral replication process, and it is believed that this protein plays a role in the rearrangement of the golgi complex membrane, facilitating the traffic of intracellular proteins (DONALDSON *et al.*, 2008); (b) nucleoside triphosphatase 2C-like (NTPase); (c) p20 or p22 (depending on the genogroup), (d) viral genome-associated protein (VPg), which binds covalently to genomic RNA and acts in various roles in the replication cycle, (e) proteinase (Pro) and (f) RNA-dependent RNA polymerase (RdRp) (JIANG *et al.*, 1993; GLASS *et al.*, 2000; BELLIOT *et al.*, 2003; HARD *et al.*, 2005; ALLEN *et al.*, 2008).

ORF-2 encodes the main capsid protein (VP1), consisting of 530 to 555 aa (JIANG, WANG, GRAHAM and ESTES, 1992; LE-PENDU *et al.*, 2006), and plays an important role in the replication and assembly of the virus. As previously mentioned, VP1 is organized into two main domains: SeP, with the S domain surrounding the viral genome, presenting more internally and linked to a third N-terminal domain, and the more external P domain, presenting an arc-shaped protrusion emanating from the surface and is involved in dimeric interactions to stabilize the capsid (GREEN *et al.*, 2007; ATMAS & ESTES, 2001; BERTOLOTTI-CIARLET *et al.*, 2003). The P domain is subdivided into two domains: P1 and P2, the former being more internal forming the base of the capsomer arches, and is implicated in the antigenicity of the virion. The P2 domain is located more externally, showing high variability, and is important in the host's immune response (figure 3) (TAN *et al.*, 2003; CAO *et al*, 2007).

ORF-3 encodes a smaller structural protein, called VP2, with one or two copies per virion located in the 3' region of the genome, which regulates the expression of VP1 and may participate in the process of encapsidation of the genomic RNA and assembly of the virion (GLASS *et* al., 2000). The VP2 protein is made up of 208 - 268 aa and shows high variability in its sequence in different genotypes (Figure 3) (HARDY *et al.*, 2005).

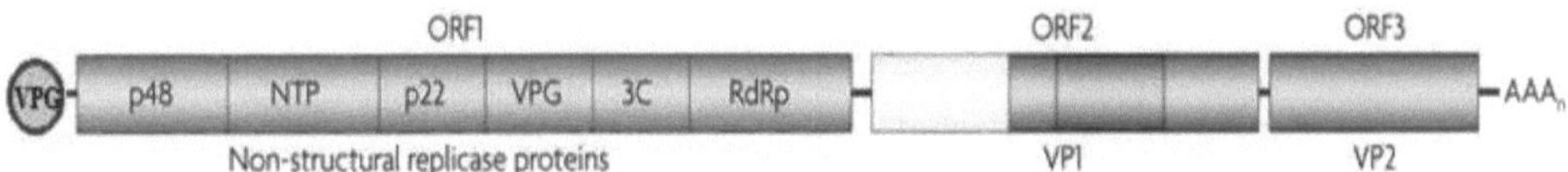

Figura 3: Representation of the norovirus genome, which is composed of three open reading regions (ORFs), which encode non-structural and structural proteins, essential for the assembly and structural organization of the virus. **Source:** Adapted from DONALDSON et al. (2010).

Based on the sequence of the gene that encodes the VP1 protein, ZHENG et al. (2006) established the classification of the norovirus genus into three levels, according to the percentage of identity of the aa: strain (0 - 14.1%), genotype (14.3 - 43.8%) and genogroup (44.9 - 61.4%). They defined five genogroups (GI, GII, GIII, GIV and GV) and 29 genotypes (8 in GI, 17 in GII, 2 in GIII, 1 in GIV and 1 in GV); Genogroups GI, GII and GIV represent human NoV, except for GII/11 (figure 4).

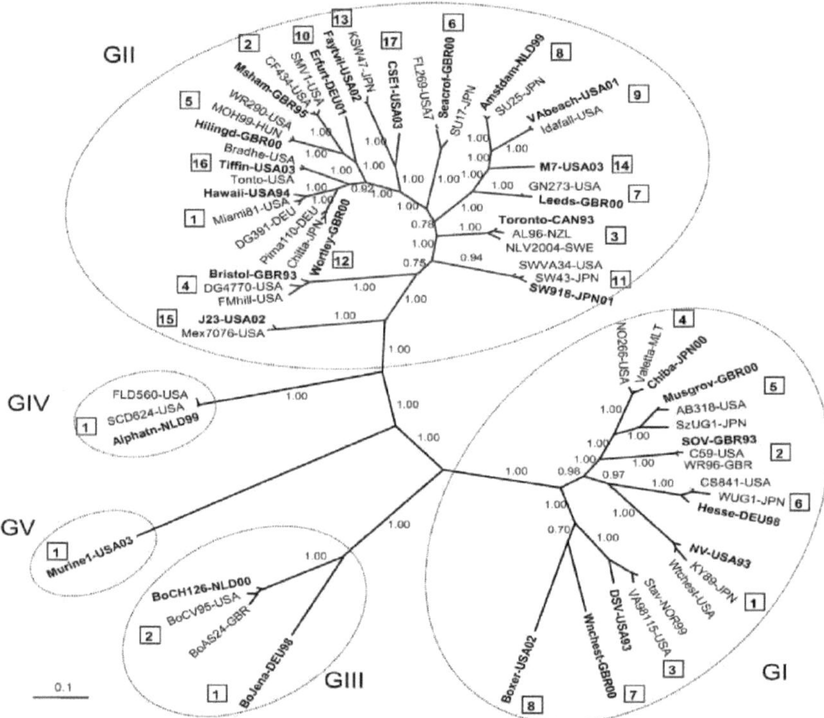

Figura 4: Phylogenetic analysis using a 141 amino acid sequence of the VP1 protein, showing the nomeclature proposed by ZHENG *et al.*, (2006) classifying NoV into five genogroups. **Source:** ZHENG *et al.*, (2006).

4.3 Biological characteristics

NoVs are very resistant under different conditions; a bacteria-free fecal filtrate containing NoVs at pH 2.7 remains infectious for 3 hours at room temperature, which explains the ability of NoVs to remain infectious after passing through the stomach (DOLIN *et al.*, 1972 ; CAUL *et al.,* 1996); they resist treatment with 20% ether for 18 hours at 4°C and show resistance to incubation for 30 minutes at 60°C. NoVs are also resistant to inactivation by treatment with chlorine at a concentration of 3.75 to 6.25 mg/L (free residual chlorine of 0.5 to 1.0 mg/L), usually found in water distributed by the water supply system, and are only inactivated after treatment with chlorine at 10 mg/L, being more resistant than Poliovirus type 1, human rotavirus and bacteriophage f2

(KESWICK *et al.,* 1985 ; GRENN et *al.,* 2001).

In recent years, some studies have suggested that the infectivity of NoV may be related to host characteristics, when looking at susceptibility and resistance to infection by these viruses. Among these characteristics is a complex of carbohydrates present in the red blood cells, urinary tract, respiratory tract and digestive tract called HBGAs (*"histo-blood group antigens"*) (LE PENDU *et al.,* 2004), which can also be found as free oligosaccharides in biological liquids such as milk, saliva, blood and intestinal contents (TAN *et al.,* 2005 ; JIANG *et al.,* 2005). These carbohydrate groups are extremely polymorphic and have three main families: Lewis, ABH and secretory, which are responsible for binding NoVs to intestinal cells (MARIONNEAU *et al.,* 2002; TAN, HEGDE, JIANG; 2004). Another factor is the expression or not of the FUT-2 gene, which codes for the $\alpha(1,2)$ fucosyltransferase enzyme that adds fucosyl groups for the synthesis of precursor chains of HBGAs, which are converted into H type1 antigenic structures, surface carbohydrates of intestinal tract epithelial cells and mucosal secretions. Studies carried out with VLPs have shown that non-secretor individuals, i.e. those who do not express the FUT-2 gene (and therefore do not have the H type 1 carbohydrate), are less susceptible to infection by the virus (LE PENDU *et al.,* 2006); but they do not have resistance to other *Caliciviruses* (LINDESMITH *et al.,* 2003).

4.4 Viral replication

NoV enter the body orally, mainly through contaminated food or water (HANSMAN *et al.,* 2004; KOOPMANS *et al.,* 2004; JIANG et *al.,* 1992; KATAYAMA et *al.,* 2006; DONALDSON *et al.,* 2008; ZINTZ et *al.,* 2005; VASICKOVA et *al.,* 2005) and survive in the acidic contents of the stomach (DOLIN et al., 1972, 1975; CAUL *et al.,* 1996). The site of primary replication has not been well defined, but it is thought that it may be the jejunal portion of the small intestine due to histopathological findings carried out on volunteers after administration of GI and GII strains in Bethesda, Maryland, USA (KARST *et al.,* 2010).

Viral adsorption occurs through the interaction between the main capsid protein VP1, through its P2 subdomain, and histo-sanguineous group antigens (HBGA) present on intestinal epithelial cells and which function as intestinal receptors for these viruses (TAN *et al.,* 2005). After binding, the particle enters the intestinal cell, followed by the release of the linear genome composed of positively polarized RNA (figure 5).

Thus, replication begins with the translation of the positive strand mediated by interactions between the VPg protein and the viral genomic RNA. The open reading region-1 (ORF-1) is initially translated to produce a non-structural polyprotein, which is then processed by the viral proteinase 3c-like (3CLpro), giving rise to non-structural proteins, including viral RNA polymerase (RpRd). Synthesis of the negative RNA strand from the positive RNA template begins at the 3' end of the positive template strand and involves interactions with viral proteins. The negative strand serves as a template for both genomic and subgenomic RNA transcription (figure 5). The large amount of positive subgenomic viral RNA serves as a template for the translation of the structural proteins VP1 and VP2 (GREEN *et al.*, 2007). Although the NoV genome has been studied since the 1990s, when the viral genome was cloned, it has not been possible to establish with certainty the entire replication mechanism of NoVs, due to the lack of an animal model, or even a cell culture mechanism for multiplication of these agents, the mechanisms of RNA packaging, maturation and release of the viral particle are still not fully understood (GREEN *et al.*, 2007).

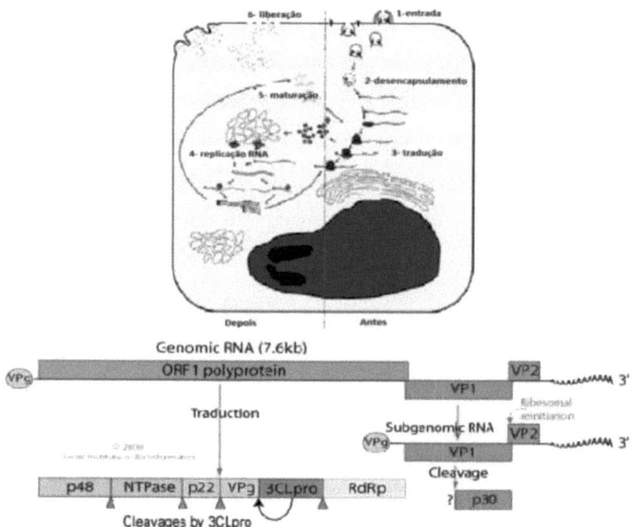

Figura 5: Diagram of the phases involved in viral replication. NoV enters the cell through the interaction between the VP1 protein and antigens of the histo-sanguineous group. Inside the cell, its ssRNA+ genome is released and directly translated by the ribosomes. In the late phase of replication, the structural proteins of the virus are translated, with subsequent assembly and release of the viral particle by the apical surface of the intestinal cell. **SOURCE:** GREEN *et al.*, (2007).

1.5 Pathogenesis and clinical manifestations

NoV are highly infectious and the ingestion of ten viral particles has an infectivity rate of 50% in inoculated patients (GRAHAM *et al.*, 1994). Studies have shown that NoV infection is self-limiting and the incubation period varies from 10 to 51 hours, with an average of 24 hours, with clinical symptoms generally lasting between 1 and 2 days (CUBITT *et al.*, 1979; DANIELS *et al.*, 2000). During experimental infection in volunteers, the excretion of viral particles detected by IME began 15 hours after inoculation at low concentrations and coincided with the peak of the disease and did not exceed 72 hours after the first symptoms, however, viral excretion has been detected for more than seven days after the onset of symptoms when analyzed by RT-PCR (GREEN *et al.,* 2007). A study of adult volunteers administered NoV-positive fecal filtrate demonstrated the excretion of viral particles for more than two weeks after the symptomatic phase (KAPLAN *et al.,* 1982a, 1982b; PARASHAR et *al.,* 1998; MURATA et *al.,* 2007), and some volunteers present asymptomatic infection with viral excretion (GRAHAM *et al.,* 1994; OKHUYSEN *et al.,* 1995). A study carried out in a pediatric clinic showed that the average time of viral excretion in children under the age of three was 16 days, ranging from 5 to 47 days (MURATA *et al.,* 2007).

After studying biopsies of intestinal tissue, it was established that the primary site of viral replication was in the jejunum. At these sites, the tissue showed histopathological lesions, with shortening of the intestinal villi, vacuolization of the cytoplasm and infiltration of the lamina propria by mononuclear cells (AGUS *et al.,* 1973; DOLIN *et al.,* 1975; CAUL *et al.,* 1996).

with intact mucosa, in volunteers who developed gastroenteric disease, after administration of bacteria-free fecal filtrate (GREEN *et al.,* 2001). A delay in gastric emptying was also observed in these individuals. It has been proposed that abnormal gastric motor function is responsible for the nausea and vomiting associated with these viral agents (ROCKX *et al.,* 2002).

Clinically, the disease is manifested by abdominal pain, with or without nausea, vomiting and/or diarrhea characterized by watery stools. The development of symptoms can be gradual or abrupt and, in general, adults present diarrhea as the predominant symptom of the disease, while nausea and vomiting are more frequently observed in children. Other clinical manifestations commonly observed are headaches, low-grade fever, chills and myalgia, occurring in 25 to 50% of infected people (THORNTON *et al.,* 2004, ATMAR *et al.,* 2008). Clinical manifestations observed in 38 outbreaks of gastroenteritis caused by NoV were: nausea (79%), vomiting (69%), diarrhea (66%), abdominal pain (30%), headache (22%), fever (37%), chills (32%), myalgia (22%), with no mucus or blood in the stool (GREEN *et al.,* 2007). The infection can worsen in children under the age of five, the elderly and immunocompromised patients (WILHELMI *et al.,* 2003; THORNTON *et al.,* 2004), 2004); in these patients, diarrhea followed by concomitant vomiting leads to severe dehydration resulting in an electrolyte imbalance, leading to hospitalization, which can become an aggravation in places where health services are inadequate or even non-existent (NAKATA *et al.,* 1998).

1.6 Laboratory diagnosis

NoV research faces some difficulties due to the low viral excretion in the feces of infected patients, and the lack of a cell culture for propagating this agent. The first time NoV was identified, electron immunomicroscopy (EMI) was used, which is more sensitive than electron microscopy (EM). EMI is based on viral morphology to distinguish between the viruses that cause gastroenteritis (KOOPMANS *et al.*, 2002) and, because it is more sensitive, requires a concentration of 104 viral particles per mL. The sample is incubated with standard-specific serum, which agglutinates the viral particles, making visualization more effective (KAPIKIAN *et al.*, 1994). Some modifications of IME have been described, such as solid phase immunomicroscopy (SPIEM) for the detection of these viruses (ATMAR & ESTES, 2001).

Immuno-adherence hemagglutination (IAHA) was developed to evaluate the level of antibodies in seroprevalence studies. It is a sensitive and specific assay, requiring a small amount of antigen. IAHA is not efficient for detecting antigen in feces due to the presence of unspecific agglutinins, requiring the use of partially purified antigen (GREENBERG; KAPIKIAN, 1978). Radioimmunoassay (RIA) has been used to replace IAHA as it is a more sensitive method that uses a smaller amount of antigen and was developed as an alternative to IME for detecting antigens in feces.

A widely used technique for detecting viral antigens in feces is the enzyme-linked immunosorbent assay (EIE), which uses hyperimmune sera obtained from laboratory animals infected with animal calicivirus and is suitable for large-scale use in epidemiological studies. However, these tests are less effective for humans due to the great genetic diversity of human NoV, requiring a panel of hyperimmune antisera for different types of HuCVs (CHIBA *et al.*, 2000). In order to develop more effective tests, it is necessary to produce more recombinant capsid antigens, representing more antigenic types, making the test more sensitive (JIANG *et al.*, 2000). These tests are currently used to detect viruses in feces, with a qualitative methodology, starting with a 1:100 dilution of the feces sample with a buffered NaCl solution containing 0.1% NaN3, and then tested on plates containing peroxidase-conjugated monoclonal antibodies, capable of identifying samples belonging to both genogroup I and genogroup II. The antigen-antibody bond is then revealed with a peroxide solution containing urea/TMB. These tests have a sensitivity of 95% and a specificity of 95.2%, which gives the test greater reliability when compared to RIA, and the advantage of not using radioisotopes (CHIBA *et al.*, 2000; JIANG *et al.*, 2000; RICHARDS *et al.*, 2004).

Since the 1990s, with the availability of molecular techniques for amplification, sequencing and genomic expression, it has been possible to characterize the genome and antigens of NoVs through different regions of the genome, such as the RdRp region located in ORF1, the ORF1 - ORF2 junction region and the N-terminal region of the VP1 capsid protein (KOBAYASHI *et al.*, 2000).

Reverse transcription polymerase chain reaction (RT-PCR) has been widely used to diagnose NoV infection by detecting its genome in clinical samples. The complete sequencing of some samples has allowed the design of genogroup-specific primer pairs for the genotyping of samples. Compared to ME, RIA and EIE, RT-PCR is a more effective method for viral detection, considering the small amount of viral material required in the specimen, as well as a longer period of viral detection post-infection (BORGES *et al.*, 2006).

The primer pairs most commonly used in RT-PCR to detect the presence of NoV in feces are those designed from the RdRp gene, as this is the most conserved region in the HuCV genome (ANDO *et al.*, 1995; GREEN *et al.*, 1995; JIANG *et al.*, 1999). The use of molecular techniques has shown a very great genetic diversity among these viruses, and a difficulty in developing a detection system that encompasses all caliciviruses. It is therefore advisable to use multiple areas of the genome as targets for primers, thus increasing the chance of viral detection (GREEN *et al.*, 1995; ATMAR & ESTES, 2001).

Real-time PCR, also used in the diagnosis of HuCVs, identifies the target RNA with greater sensitivity, since the amplification is detected through fluorecence capture. This technique can also have greater specificity due to the use of a specific probe for the target fragment in the reaction. DNA amplification and detection are carried out simultaneously in a closed system, eliminating the need for additional procedures such as electrophoretic running of the products on agarose gels. By eliminating these steps, results are obtained earlier with real-time PCR than with conventional PCR. It is an automated method that allows quantitative analysis (RICHARDS *et al.*, 2004).

1.7 Epidemiology

NoVs infect people of all ages, a characteristic that distinguishes them from other viruses that cause gastroenteritis, such as RV, AdEV and AsTV, which preferentially infect children up to five years of age (GLASS *et al.*, 2000). They are considered to be the main causative viruses of this disease, accounting for approximately 80% of cases of viral gastroenteritis, causing outbreaks in various environments such as schools, hospitals, nursing homes, nurseries, military barracks and ships (FANKHAUSER *et al.*, 2002; GALLIMORE et *al.*, 2004; VIDAL *et al.*, 2005), 2005) The mode of transmission of these viruses is predominantly by the fecal-oral route, by ingestion of contaminated water and food, and by person-to-person contact (figure 6), or even by aerosols produced during vomiting (MARKS et al., 2000; PANG *et al.*, 2000 ; MORENO-ESPINOSA *et al.*, 2004).

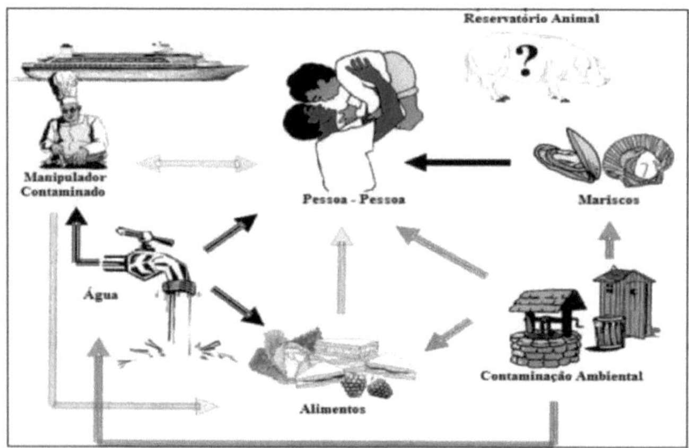

Figure 6: Main routes of transmission. NoV is transmitted predominantly by the fecal-oral route through the ingestion of contaminated water and food and by person-to-person contact. **Source:** Moreno-Espinosa *et al.* (2004)

Contamination of food and water used for consumption or recreational activities can serve as a primary source of outbreaks, since NoV are very stable and resistant to water treatment with chlorine, and can remain infectious for long periods in these environments (PARASHAR; MONROE, 2001). Currently, NoV are considered to be the most important etiological agents causing outbreaks of non-bacterial gastroenteritis, especially in closed environments (VINJÉ *et al.*, 1996; DEDMAN et al., 1998; INOUYE *et al.*, 2000). Epidemic outbreaks usually occur in nurseries, schools, nursing homes, cruise ships, infirmaries, restaurants and social events; they can vary in the number of people involved, from small family groups to thousands of individuals, as occurred in an outbreak in Japan involving 5,000 children (MATSUNO *et al.*, 1997 ; GREEN *et al.*, 2007). In studies carried out in 33 US states, FRANKHAUSER *et al.* (1998) defined the epidemiological characteristics of 90 outbreaks of non-bacterial gastroenteritis in different environments. It was observed that outbreaks were more common in nursing homes and restaurants. The most common mode of transmission was from food, water and direct person-to-person contact, results similar to those found in other parts of the world (KOOPMANS *et al.*, 2002).

The role of food in the epidemiology of infections has been demonstrated by the increase in outbreaks related to the consumption of contaminated food. NoV are highly infectious and can cause disease with as few as 10 viral particles. The high attack rate, the high number of asymptomatic carriers, the prolonged period of viral excretion and the high resistance of viruses outside the host contribute to the explosive nature of food-borne NoV outbreaks (CAUL; APPLETON, 1982; GREEN *et al.*, 2001).

Most individuals in developed and developing countries show evidence of NoV infection before adulthood, reflecting the global distribution and edemic nature of these viruses. Studies with enzyme-linked immunosorbent assays using NoV VLPs to test for antibodies against genogroup I and II (G1 and G2) have shown that the prevalence of antibodies to each genogroup gradually increases with age, especially in developed countries. The prevalence of antibodies against GII is higher when compared to GI in most studies, reflecting the predominance of circulating genogroup II strains (NOEL *et al.*, 1997; GREEN, 2007). The GII/4 genogroup has been the most frequent genotype detected in outbreaks of acute diarrhea caused by NoV in the world, showing its predominance and global distribution (FANKHAUSER et al., 1998; HALE *et al.*, 2000; KOOPMANS *et al.*, 2001; LOPMAN et *al.*, 2002).

In Brazil, the first study of NoV prevalence was described by PARKS et al. in 1999, in hospitalized children with or without diarrhea; however, in 1993 TIMENESKY et al. had already reported NoV observations in children's stools in Sao Paulo - SP. Seroprevalence studies described in some Brazilian regions revealed a range of 39% to 100% positivity for antibodies against NoV in eight indigenous communities (GABBAY *et al.*, 1994); and 71% in children in Cearà (TALAL *et al.*, 2000). The prevalence of this viral agent increases with age and can reach 90% in adults in several countries (CARDOSO *et al.*, 2005). Studies carried out on outbreaks in the state of Sao Paulo showed that the NoV positivity rate was 15.7% and 29.2%, as published by Morillo et al. (2008) and Cilli *et al.* (2011) respectively, showing that the prevalence of this infectious agent has increased in this region. Other regions studied showed rates ranging from 9% to 35%; such as 18% in Belém do Parà (ARAGÂO et al., 2010), 13.2% in Porto Alegre, RS (GEORGIADIS *et al.*, 2010), 23% in Florianópolis, SC (VICTORIA *et al.*, 2009), 9% in Salvador, BA (XAVIER *et al.*, 2009) and 35% in Rio de Janeiro, RJ (FERREIRA *et al.*, 2010), which can be explained by the high flow of tourists in this region. In all the outbreaks studied, the predominant genotype was GII, which corroborates data found worldwide (NOEL *et al.*, 1997; GREEN, 2007; LINDESMITH *et al.*, 2008).

In other countries, seroprevalence of antibodies against NoV has been detected at all ages, with levels of over 70%, such as in the USA, Switzerland and Ecuador (GREENBERG *et al.*, 1978a, 1978b). Zintz *et al.* (2005) observed a prevalence of 7.1% in children admitted to three children's hospitals in the USA. In Europe, Lopman et al. (2002) found an overall NoV positivity rate of 11%, but with countries with lower rates such as England (6%) and higher rates such as Finland (20%). In Asia, NoV has been detected in two regions of Thailand, with positivity rates of 12% (HANSMAN *et al.*, 2004) and 44.7% (KITTIGUL *et al.*, 2010), and 15% in Korea (Yoon *et al.*, 2008). These data together show that infections caused by NoV occur all over the world.

1.8 Immunity and Vaccination

The host's immune response to NoV is not well understood due to the lack of an animal model, or even the lack of a form of cell culture where the production of antibodies by specialized cells could be quantified by an assay. What little is known comes from studies on volunteers challenged with the virus, and the response pattern observed does not seem to resemble that of other viruses (GREEN *et al.,* 2007). These studies in volunteers reported two patterns of response: short-term and long-term. Short-term immunity lasted between 6 and 14 weeks; during this period, individuals were challenged with viruses of the same serotype and did not develop the disease; when challenged with viruses of different serotypes, they developed the disease; which suggests that short-term immunity may be serotype-specific. In long-term immunity, the volunteers showed two response patterns; of twelve volunteers tested, six developed the disease both at the initial challenge and 27 and 42 months after the first challenge, and six patients did not develop the disease in any of the challenges. When serum antibodies were tested by IME and RIA before reinfection in the twelve volunteers, titers were higher in individuals who developed the disease, and volunteers who did not become ill had little or no antibodies after the initial challenge, and did not develop an immune response after any challenge (GREEN *et al.,* 2005, 2007). This may be explained by the association between NoV infection and specific HBGAs, which relates to host susceptibility factors; it is now known that patients who did not become infected through repeated exposure to NoV did not have specific receptors for the virus in the intestinal mucosa (TAN & JIANG, 2007).

Studies of the viral genome and structural knowledge of NoV, especially the viral proteins, have provided advances that support the development of a vaccine using recombinant DNA technology, based on VLPs that demonstrate immune responses in humans (TACKET *et al.,* 2003). The P-domain of the main capsid protein (VP1), which is responsible for viral interaction with the host cell, is promising for vaccine development, as it is immunogenic and easily produced by gene recombination using *Escherichia coli* (TAN & JIANG, 2007; TAN *et al.,* 2009). The difficulty in establishing an effective vaccine against NoV is limited by the high capacity of these agents to undergo mutations in ORF2, the region that encodes the VP1 protein, where the P-domain region is located, involved in the antigenicity of the virus, which favors escape from the immune system, which makes further studies necessary in order to develop an effective vaccine for the prevention of these infections (PATEL *et al*, 2008).

1.9 Prevention and control

In outbreaks of acute gastroenteritis, interrupting transmission by eliminating the source of infection is the main strategy for control, especially in closed environments where most outbreaks attributed to NoV occur. Attention to hand hygiene after contact with patients or objects that may be contaminated is essential (WILHELMI *et al.,* 2003). NoV can be transferred by direct contact through individuals with contaminated hands to up to seven different surfaces, such as objects for personal or collective use, with direct implications for the origin of outbreaks and the spread of infection. The use of detergents is ineffective in eliminating the contamination of these viruses, requiring the use of a hypochlorite/detergent solution to decontaminate surfaces (BARKER *et al,* 2004).

The vaccine is a preventive tool that could be used to reduce morbidity, since sanitation measures are not enough to prevent the occurrence of acute NoV gastroenteritis. VLPs were used to immunize mice and volunteers, and immunogenicity was assessed by the presence of antibodies in serum and faeces. The oral formulations of VLP expressed in baculovirus and in transgenic plants used in these studies produced a mucosal humoral response in the challenged mice. These results represent the first advances towards obtaining an anti-NoV vaccine in the future (ESTES *et al.,* 2000 ; NICOLIER-JAMOT *et al.,* 2004 ; TACKET *et al.,* 2005).

2 BACKGROUND

Acute diarrheal disease is one of the world's biggest public health problems, both in developed and developing countries, the latter being a major cause of morbidity and mortality, especially in children under five years of age. Acute diarrheal disease can be caused by viruses, bacteria and protozoa, with RV, NoV, AdE, AsT and SaV being responsible for a large proportion of these diseases. NoVs are the most important causative agents of diarrheal outbreaks on all continents, affecting individuals of all ages, with an increasing rate in the last 8 years, when the vaccine against RVs became available, which put NoVs in the spotlight. There is little data on the number of cases occurring in some Brazilian regions, which can be explained by the lack of a low-cost diagnostic method and information on the importance of this agent. The few epidemiological studies that have been published in Brazil come from research centers, allowing us to elucidate the prevalence of these agents in various regions. The Triângulo Mineiro region is crossed by a vast highway network that connects the main Brazilian capitals, resulting in a large flow of people from other states and regions, which can be crucial for the spread of viral agents. As such, this study is of great importance not only to elucidate epidemiological data and the genetic identity of NoV in this region, but also to serve as a source of data for future prevention and control program initiatives that may be developed and implemented in the country.

3   GENERAL OBJECTIVE

To detect and characterize NoV strains in the feces of children with symptoms of acute infectious diarrhea in the cities of Uberaba and Uberlândia, located in the Triângulo Mineiro region.

2.1 Specific objectives

I-Determine the prevalence of norovirus infection by detecting the virus in feces.

2-Identify the genogroups of the norovirus detected;

3-Analyze the genetic diversity of the noroviruses detected through partial genomic sequencing and phylogenetic analysis;

4- Analyze the epidemiological aspects observed in the cases with regard to their association with the molecular characterization of the noroviruses found.

4 MATERIAL AND METHODS

4.1 Study population and samples

This was a prospective study to analyze the prevalence and characterization of NoVs in children treated in the cities of Uberaba and Uberlândia, Minas Gerais. This study was approved by the ethics committee of the Federal University of the Triângulo Mineiro (CEP/UFTM) under protocol number 672 (Annex 1).

The population studied was made up of children aged between zero and ten with clinical symptoms of acute infectious gastroenteritis, treated in private and public hospitals and clinics. The samples were obtained from the Jorge Furtado (JF) laboratory in Uberaba and the IPAC Medicina Diagnóstica (IPAC) laboratory in Uberlândia. Detection of rotavirus was requested for all samples. Together with the samples, an identification form was sent containing the children's data such as gender, age, date of collection and the result of the Látex test for RV. The samples were then recorded, labeled and kept at -20 °C until they were analyzed for NoV.

4.2 Detection of NoV in fecal samples by EIE

The samples were screened for the presence of NoV by enzyme-linked immunosorbent assay (ELISA), using the *RIDASCREEN® Norovirus 3rd Generation* diagnostic kit (R Biopharm AG Landwenrstr, Germany) according to the manufacturer's instructions. This test consists of the qualitative detection of NoV genogroups I and II in stool samples using a "sandwich" methodology and monoclonal antibodies. Initially, a 1:10 dilution of the fecal samples was prepared in 1 ml of diluent 1 (annex 2, item 1). This dilution was left to stand for 10 minutes at room temperature and then centrifuged for 5 minutes at 2500xg. The supernatant was used in the ELISA to detect NoV.

Before starting the test, all the reagents and the microplate were removed from the refrigerator and placed at room temperature until they reached 2025°C. At the same time, the washing solution was prepared according to the manufacturer's instructions (Appendix 2, item 1). The microplates were removed from the protective aluminum packaging and 100µl of the positive control was pipetted into cavity A1, 100µl of the negative control into cavity A2, and 100µl of each supernatant of the previously diluted stool samples into the other cavities of the microplate. This was followed by pipetting 100µl of the 1

(biotin-conjugated anti-norovirus antibodies) to each well of the plate. The microplate containing the controls, samples and conjugate 1 was manually homogenized for 2 minutes, protected from light by a sticker attached to the top of the microplate, and incubated at room temperature for 60 minutes. After the first incubation, the microplate was washed five times with the washing solution described above using a microplate washer, using 300µl of the solution per well per wash cycle. At the end of the five wash cycles, the microplate was poured onto filter paper in order to remove any remnants of the wash solution contained inside the cavities. Subsequently, 100µl of conjugate 2 (streptavidin-peroxidase conjugate) was added to each cavity of the plate, which was again protected from light by an adhesive fixed to the top and then incubated for 30 minutes at room temperature. The microplate was washed again and poured onto filter paper as above. Subsequently, 100µl of the substrate was added to each plate cavity (hydrogen peroxide/TMB), which was incubated for 15 minutes in a dark environment at room temperature and then the reaction was blocked by adding 100µl of stop solution (Stop - 1N sulfuric acid) to all the plate cavities, which was then homogenized for 2 minutes and the test was read on a spectrophotometer at an absorbance of 450nm. To determine the results, the *cut-off* point was calculated according to the kit's specifications, using 0.15 of the absorbance unit added to the absorbance of cavity A2 (negative control). Samples were therefore considered positive (presence of NoV) when their absorbance was 10% higher than the *cut-off* value, while samples with an absorbance 10% below the *cut-off* value were considered negative (absence of NoV). Samples whose readings fell between the values of the positive and negative samples were classified as indeterminate and in these cases the whole procedure was repeated.

4.3 Nucleic acid extraction from fecal samples

The fecal samples received during the study were thawed to prepare a fecal dilution, the (clarified) supernatants of which were used for nucleic acid extraction. In sterile 1.5 ml microtubes, the feces were diluted to 20% in phosphate-buffered saline (Annex 2, item 2). The samples were then centrifuged at 14000 rpm for 15 minutes and 300µl of the supernatant was used for the extraction of viral RNA using the method of Boom *et al.* (1990), with modifications suggested by Cardoso *et al.* (2002). This extraction is based on the lysis and nuclease inactivation properties of guanidine isothiocyanate (GITC) and also on the binding properties of nucleic acids to silica particles.

To each microtube, 300µl of clarification, 800µl of L6 cap and 20µl of Proteinase K (Invitrogen®/Life Technologies) were added. Homogenization was carried out by inversion and incubation in a water bath at 56°C for 10 minutes. Then 200µl of ice-cold absolute ethanol and 20µl of silica solution (*Sigma-Aldrich®*) were added; the samples were homogenized in a vortex mixer for 15 seconds and incubated at room temperature for 15 minutes under constant agitation. After this period, the samples were centrifuged at 9000xg for 15 seconds and the supernatant was discarded by inversion. 500µl of L2 cap was added to the pellet (viral RNA bound to the silica particle) and the microtubes were then homogenized in a vortex mixer for 15 seconds to resuspend and wash the *pellet* and centrifuged again for 15 seconds at 9000xg. 500µl of ice-cold 70% ethanol was added to the *pellet*, homogenized in a vortex mixer for 15 seconds, centrifuged at 9000xg for

15 seconds and the supernatant discarded by inversion. Next, 500µl of ice-cold P.A. acetone was added, and the microtube was manually poured for 15 seconds, again centrifuged at 9000xg for 15 seconds and then the supernatant was discarded. The microtube was dried in a water bath at 56°C for 10 minutes, with the lid open, and then 40µl of sterile distilled water was added to elute the RNA; the microtube was taken to a water bath at 56°C and incubated for 15 minutes. The microtube was then homogenized in a vortex mixer for 15 seconds, centrifuged at 9000xg for 3 minutes and 35 µl of the supernatant (sterile distilled water and RNA) were collected in a new sterile microtube and stored at -20°C.

4.4 Reverse transcription reaction (RT)

In a 200µl microtube, 16µl of the previously extracted RNA was added along with 1µ of the random primers (Invitrogen) at a concentration of 300ng/µl. The tube was subjected to a temperature of 70°C for 10 minutes to denature the RNA and immediately taken to an ice bath for 15 minutes. After denaturation, 15µl of the random RNA/initiator mixture was then added to a reverse transcription reaction mix (Table 1 in Appendix 2) to synthesize the cDNA. The reaction was incubated at 25°C for 10 minutes for the *random primer to* hybridize to the RNA, followed by incubation at 37°C for 50 minutes for the reverse transcriptase to extend, and 70°C for 15 minutes to inactivate the enzyme.

4.5 Genotyping by polymerase chain reaction (PCR)

In this reaction, 10µ of cDNA was used to characterize the viral genogroup present in the sample. Two pairs of primers were used, G1SKF/G1SKR, which identify genogroup I, and COG2F/G2SKR, which identify genogroup II (KOJIMA et al., 2002; SHINOHARA et al.,2002). When the results were inconclusive, another five pairs of primers were used, CAL-32/MO3-N and MON431/MON433 which identify genogroup I, and JV-12/ACAL-36, MON432/MON434 and MON381/MON383 which identify genogroup II (VINJÉ et al., 1997 ; BEURET et al., 2002 ; NOEL, et al., 1997). All the *primers* and the regions they amplify are shown in table 1 and figure 7.

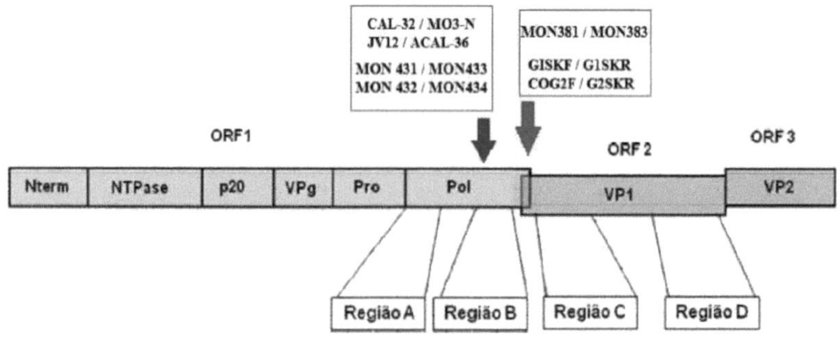

Figure 7: Diagram of the NoV genome showing the hybridization sites of the *primers* used. *Primers* used for viral detection (region B) and genotyping (region C) amplifying the partial sequences of the genes coding for the polymerase and the VP1 protein, respectively.

10µl of cDNA were added to a 200µl microtube containing 40µL of the PCR reaction mixture (Table 2 in Appendix 2). The microtube was subjected to the following conditions: 94°C for 2 minutes; followed by 35 cycles at 94°C for 30 seconds, 55°C for 30 seconds, 72°C for 60 seconds; final extension at 72°C for 7 minutes and cooling to 4°C.

Table 1: Description of the *primers* used for the detection and genotyping of noroviruses in fecal samples

Genogroup	*Primer*	Sequence (5' → 3')	Polarity	Location .	Amplicon
G1	GISKF	CTG CCC GAA TTY GTA AAT GA	*Foward*	5342-5361	330pb
	GISKR	CCA ACC CAR CCA TTR TAC A	*Reverse*	5671-5653	
G2	G2SKR	CCR CCN GCA TRH CCR TTR TAC AT	*Reverse*	5401-5423	387pb
	COG2F	CAR GAR BCN ATG TTY AGR TGG ATG AG	*Forward*	5003-5028	
G1	CAL-32	ATG AAT ATG AAT GAG GAT GG	*Forward*	4490-5127	638pb
	MO3-N	TCA GAT GGG TCT TCA TGA TTG G	*Reverse*	4490-5127	
G2	JV-12	ATA CCA CTA TGA TGC AGA TTA	*Forward*	4552-4572	428pb
	ACAL-36	GAC AAA ACA GAA GGA CCA AT	*Reverse*	4961-4980	
G2	MON 381	CCA GAA TGT ACA ATG GTT ATG C	*Forward*	5362-5383	319pb
	MON 383	CAA GAG ACT GTC AAG ACA TCA TC	*Reverse*	5661-5383	
G1	MON 431	TGG ACI AG(A/G) GGI CC(C/T) AA(C/T) CA	*Forward*	5285-5305	319pb
	MON 433	GAA (C/T)CT CAT CCA (C/T)G AAC AT	*Reverse*	5285-5305	
G2	MON 432	TGG ACI CG(C/T) GGI CC(C/T) AA(C/T) CA	*Forward*	5285-5305	319pb
	MON 434	GAA (C/G)C GCA TCC A(A/G)C GGA ACA T	*Reverse*	5285-5305	

4.6 1.5% agarose gel electrophoresis

The PCR product was subjected to electrophoresis on a 1.5% agarose gel in 0.5X Tris-Borate-EDTA (TBE) buffer (Appendix 2). 6µl of amplified DNA mixed with 2µl of *loading buffer was* pipetted into each *slot*. Electrophoresis was carried out at 100V for 45 minutes and visualization after soaking in a 0.5µg/mL ethidium bromide solution was carried out using a UV transluminator at a wavelength of 312 nm. In parallel, a 100pb DNA molecular weight standard (Invitrogen®) was used to define fragment sizes.

4.7 Purification and quantification of RT-PCR amplified products

After electrophoresis, the products were extracted from the agarose gel using a nucleic acid purification kit (*Wizard SV* Gel and PCR Clean-Up System from Promega®). The concentration of the purified DNA fragments was estimated by running them on a 1.5% agarose gel together with a Low DNA Mass Ladder concentration indicator marker (Invitrogen®).

4.8 Sequencing and purification reaction

The purified PCR products were sequenced using the *BigDye Terminator Cycle Sequence kit 3.1* on the ABI 3130 Genetic Analyzer sequencer (*Applied Biosystems®)* with the same PCR *primers.* The sequencing reaction was based on the protocol of Platt et al., 2007, with some modifications; the mixture used was described in (table 3 in appendix2). The plate was taken to the thermal cycler and subjected to the program: 95°C for 1 min, 20 cycles of 95°C for 15 sec; 55°C for 10 sec; 60°C for 2 min, followed by 10 PCR cycles (95°C for 15s ; 55°C for 10s ; 60°C for 3 min; at the end the temperature was reduced to 4°C.

The products of the sequencing reaction were purified using precipitation with 1μL of 125mM EDTA, 1μl of 3M sodium acetate (pH 5.2) and 25μl of ice-cold 100% ethanol. The plate was sealed, mixed inverted and incubated for 15 minutes at room temperature, away from light. It was then centrifuged at 3,000xg for 30 minutes. Immediately after centrifugation, the plate was inverted, discarding the supernatant and given a 180xg *spin* for 1 minute. Afterwards, 70% ethanol was added to each of the wells and the plate was centrifuged at 1650xg for 15 minutes at 4°C. It was inverted again, discarding its contents, and centrifuged again at 180 x g for 1 minute. Finally, the plate was incubated at 52°C for 15 minutes, covered with aluminium foil and taken to the sequencer, after which the DNA was resuspended with 10μl of Hi-Di formamide (*Applied Biosystems®)* and heated to 95°C for 1 minute.

4.9 Nucleotide sequence analysis

All the sequences obtained by the sequencing reaction were analyzed, edited and their respective consensus sequences were generated using the *Geneious* 6.0.5 program (DRUMMOND, 2011). The consensus sequences were then aligned with standard virus sequences previously published by Pang *et al.* in 2010 and reference samples taken from the *GenBank* database (http://www.ncbi.nlm.nih.gov/genbank/) using the MEGA5 program (TAMURA *et al.*, 2012). The alignments were analyzed by Bayesian inference in the MrBayes software (SCHWARZ, 1978) after choosing the best evolutionary model, estimated by the JModelTest software (POSADA *et al.*,

2008). The accession numbers of the sequences used for comparison in the alignment and for assembling the phylogenetic tree were: Genogroup I - M87661 (NLV Texas); Genogroup II - EF684915 (Shellharbour), JX445162 (AlbertaEI425), EU310927 (Houston), GU445325 (New Orleans), AF145896 (Camberwell virus), X86557 (Lordsdale virus), JF697292 (IPH143-09VG2), U02030 (Toronto virus), AF414410 (Miami), EF126966 (Nijmegen, 2006b), AY502023 (Farmington Hills, 2002), EF126964 (Terneuzen, 2006a), AB434770 (OC07138, 2008a), AB445395 (Apeldoorn 2008b), DQ78794 (Hunter, 2004), AJ004864 (Farmington hills, 1996) (Pang *et al.*, 2010).

5 RESULTS

5.1 Detection of norovirus by enzyme-linked immunosorbent assay (ELISA)

In this study, 975 samples were analyzed; of these, 446 (45.7%) were collected in Uberaba and 529 (54.3%) in Uberlândia. The overall percentage of norovirus positivity was 14.7% (144/975); analyzing the two locations separately, Uberaba had a prevalence of 16.8% (75/446) and Uberlândia, 13.0% (69/529). The study was carried out between 2006 and 2011; in Uberaba the samples were collected between 2006 and 2009, and in Uberlândia between 2008 and 2011. The distribution of samples over the study period and the annual prevalence of NoV positivity in each location are shown in Table 2.

The distribution of samples analyzed and NoV positivity in each month throughout the period studied can be seen in figures 8 and 9, which represent Uberaba and Uberlândia, respectively. In Uberaba, in 2006 and 2007, there was an increase in the number of positive samples from May onwards, with a peak in positivity between August and September and a gradual drop from October onwards; in 2008 and 2009, the positive samples were distributed over the months without peaks being observed, with no characteristic seasonality. In Uberaba, there has also been a decrease in the number of samples from diarrhea cases over the years, but it can be seen that positivity remained approximately the same between 2006, 2007 and 2008, with a sharp reduction in 2009 (figure 8).

In Uberlândia, there was a gradual increase in the annual prevalence of NoV between 2008, 2009 and 2010, corresponding to 4.7%, 13.0% and 20.0%, respectively. A decrease in prevalence was observed in 2011 (11.5%); however, there was no characteristic seasonal pattern in positivity (figure 9).

As for the number of samples analyzed in the two locations, peaks could be observed in all the years, but without a marked seasonality.

Table 2: Distribution of the samples analyzed in the study.

Patient sample	2006		2007		2008		2009		2010		2011		2006-2011	
	Analyzed(%)	Positive	Analyzed(%)	Positive	Analyzed(%)	Positive	Analyzed(%)	Positive	Analyzed(%)	Positive	Analyzed(%)	Positive	Total(%)	Positive
Uberaba	232	41 (17.6)	98	16 (16.3)	78	15 (19.2)	38	3 (7.9)					446	75 (16.8)
UberlBndia					105	5 (4.7)	161	21 (13)	150	30 (20)	113	13 (11.5)	529	69 (13)

Figure 8: Distribution of samples analyzed and samples positive for NoV over the period studied in Uberaba-MG

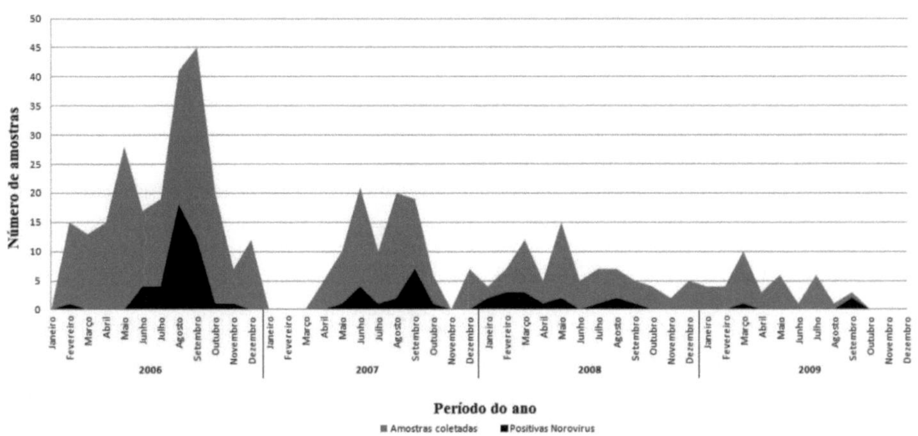

Figure 9: Distribution of samples analyzed and samples positive for NoV over the period studied in Uberlândia-MG.

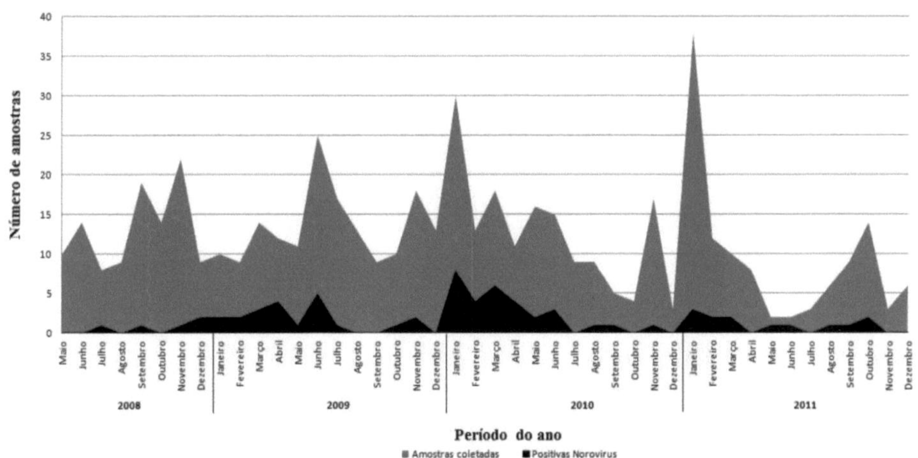

For a broader analysis of seasonality, the years studied were overlaid, and it was thus observed that, in Uberaba, the highest relative prevalences of norovirus infection occurred in the months of January (35.3%), August (31.9%) and September (30.5%); in the other months, the prevalences varied greatly, not exceeding 17.8% (figure 10). In Uberlândia, the highest prevalences occurred in the months of March (26.1%), April (25.8%) and February (23.5%); among the other months, the prevalences were also quite varied, not exceeding 16.6% (figure 11).

Tables 4 and 5 show the distribution of samples in relation to age groups over the years of the study, representing cases from Uberaba and Uberlândia, respectively. In both locations, the relative percentages of the prevalence of diarrhea and norovirus infection were higher in the 7 to 12 month age group, with 54.6% (41/75) of the samples being positive in Uberaba (table 3) and 63.7% (44/69) in Uberlândia (table 4).

We analyzed the number of samples received and the number of positive samples in relation to age groups, overlapping all the years of study in Uberaba and Uberlândia. We found that the 7-12 month age group was the most affected, accounting for 44.2% (431/975) of diarrheal samples (Figure 12) and 59.0% (85/144) of positive samples (Figure 13), respectively.

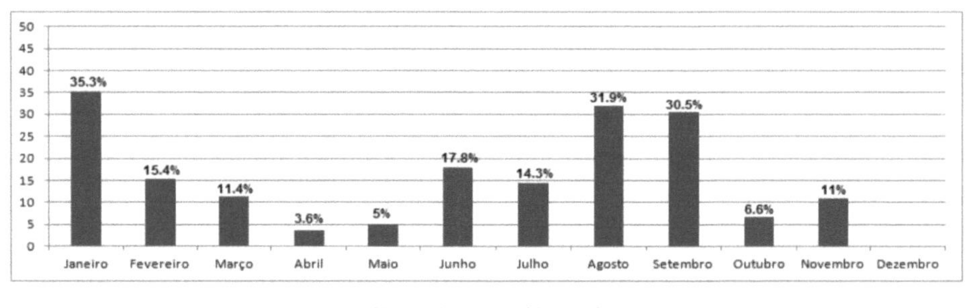

Figure 10: Percentage of positive samples distributed over the months, during all the years studied in Uberaba, MG

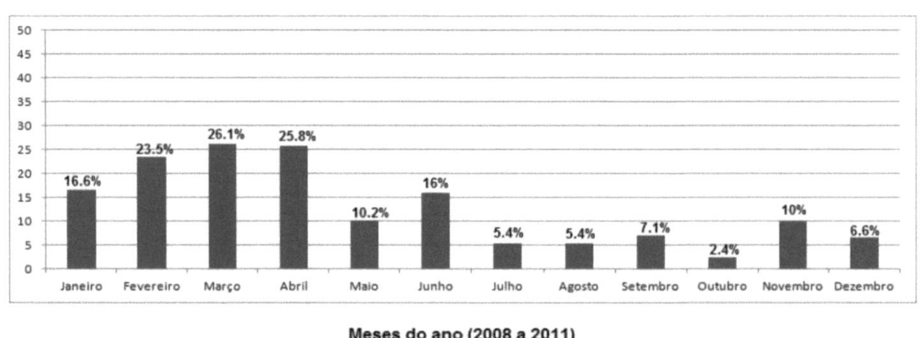

Figure 11: Percentage of positive samples distributed over the months, during all the years studied in Uberlândia, MG

Table 3: Distribution of samples analyzed and samples positive for norovirus in relation to age groups in all the years studied, in Uberaba, MG.

Age group (months)	Uberaba 2006		Uberaba 2007		Uberaba 2008		Uberaba 2009		Total	
	Analyzed	Positive (%)	Analyzed	Positive (%)	Analyzed	Positive (%) A	nalyzed	Positive (%) /	Analyzed	Positive (%)
0 - 6	31	3 (9.6)	13	3 (23)	14	0	8	0	66	6 (9)
7 - 12	100	19 (19)	36	9 (25)	29	10 (3.4)	14	3 (21.4)	179	41 (23)
13 - 24	34	7 (20.6)	12	1 (8.3)	11	4 (38.3)	1	0	58	12 (20.7)
25 - 36	17	4 (23.5)	10	0	9	1 (11)	3	0	39	5 (12.8)
37 - 48	20	2 (10)	9	1 (11)	11	0	1	0	41	3 (7.3)
>48	27	5 (18.5)	16	2 (12.5)	4	0	10	0	57	7 (12.3)
ND	3	1 (33.3)	2	0	0	0	1	0	6	1 (16.6)
Total	232	41 (17.6)	98	16 (16.3)	78	15 (19.2)	38	3 (7.9)	446	75 (16.8)

Table 4: Distribution of samples analyzed and samples positive for norovirus in relation to age groups in all the years studied, in Uberlândia, MG.

Age group (months)	Uberlândia 2008		Uberlândia 2009		Uberlândia 2010		Uberlândia 2011		Total	
	Analyzed	Positive (%)	Analyzed	Positive (%)	Analyzed	Positive (%)	Analyzed	Positive (%)	Analyzed	Positive (%)
0 - 6	9	0	14	4 (28.5)	30	2 (6.6)	14	1 (7.1)	67	7 (10.4)
7 - 12	37	3 (8.1)	84	13 (15.5)	77	19 (24.7)	54	9 (16.6)	252	44 (17.4)
13 - 24	23	1 (4.3)	29	1 (3.4)	26	6 (23)	19	2 (10.5)	97	10 (10.3)
25 - 36	19	1 (5.3)	14	1 (7.1)	7	0	12	0	52	2 (3.8)
37 - 48	10	0	10	1 (10)	6	3 (50)	8	1 (12.5)	34	5 (14.7)
>48	7	0	10	1 (10)	4	0	6	0	27	1 (3.7)
Total	105	5 (4.8)	161	21 (13)	150	31 (20.6)	113	13 (11.5)	529	69 (13)

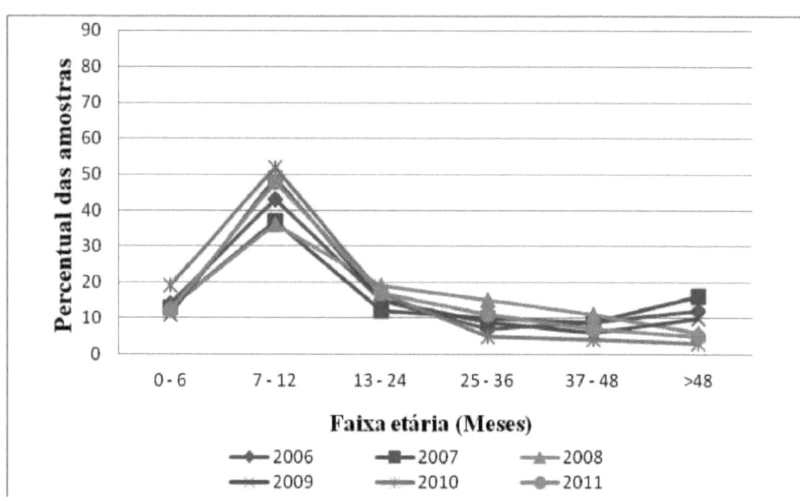

Figura 12 Percentage of diarrheal samples in relation to age groups, overlapping all years of study, in Uberaba and Uberlândia.

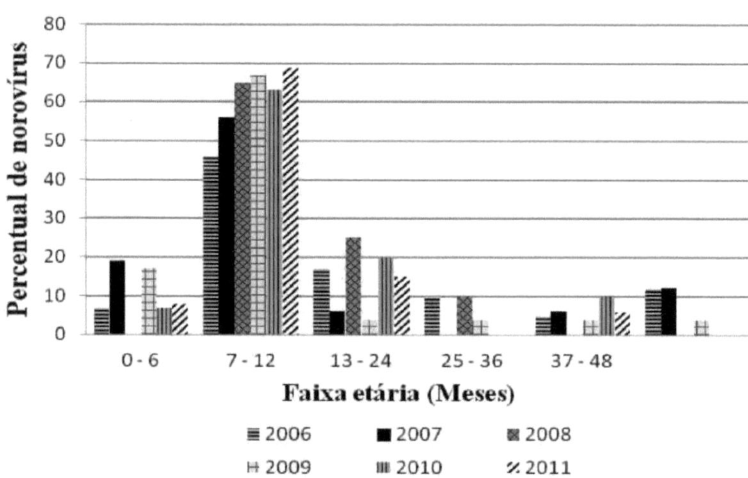

Figura 13 Percentage of samples positive for NoV in relation to age groups, in all years of study, in Uberaba and Uberlândia.

The number of samples analyzed from children treated at outpatient clinics was proportionally higher than the number of samples from hospitalized children, both in Uberaba and Uberlândia; however, when we looked at norovirus positivity, the samples from hospitalized children showed a slightly higher percentage of relative positivity (table 5).

Table 5: Origin of diarrheal samples and samples positive for NoV in Uberaba and Uberlândia, MG.

Samples	Diarrhea	Norovirus (%)
Inpatients	143 (14,7%)	26 (18,1%)
Outpatient	832 (85,3%)	118 (14,1%)
Total	975	144

Analyzing the two locations separately and for each year, it was observed that in Uberaba, only in 2007, the relative prevalence of positivity was significantly higher in samples from hospitalized children; in the other years, it was higher in samples from outpatients (table 6). In Uberlândia, it was observed that in 2007 and 2009, the relative percentages of positivity for norovirus in samples from hospitalized children were higher in 2007 (22.6%); in 2010 and 2011, samples from outpatients had higher relative percentages of positivity (table 7).

The distribution of samples in relation to age groups and origin can be seen in Table 8. In general, it was observed that in all age groups most of the samples represented outpatient care; however, when we looked at the relative percentage of positivity for norovirus in the samples from hospitalized children, the first two age groups, especially in the 0 to 6 month age group, showed higher percentages.

Table 6: Samples analyzed and samples positive for NoV in relation to the origin of the samples, in the city of Uberaba-MG

Sample of patients	Uberaba 2006		Uberaba 2007		Uberaba 2008		Uberaba 2009		Uberaba 2006-2009	
	Analyzed	Positive (%)	Analyzed	Positive (%)	Analyzed	Positive (%)	Analyzed	Positive (%)	Analyzed	Positive (%)
Inpatients	183	(16.6)	6214	(22.6)	60		60		9217	(18.4)
Outpatient	21438	(17.7)	362	(5.5)	7215	(21)	323	(9.3)	35458	(16.3)

Table 7: Samples analyzed and samples positive for NoV in relation to the origin of the samples, in the city of Uberlândia-MG

Sample of patients	Uberlândia 2008		Uberlândia 2009		Uberlândia 2010		Uberlândia 2011		Uberlândia 2008-2011	
	Analyzed	Positive (%)	Analyzed	Positive (%)	Analyzed	Positive (%)	Analyzed	Positive (%)		

	Analyzed Positive (%)				
Inpatients	141 (7.1)	174 (23.5)	153 (18.7)	40	508 (16)
Outpatient	914 (4.4)	14417 (11.8)	13527 (20)	10913 (12)	47961 (12.7)

Table 8: Distribution of the number of samples analyzed and samples positive for NoV, separated by origin and age group, representing Uberaba and Uberlândia.

Age group (months)	Origin of the samples	Analyzed	Positive (%)
0 - 6	Inpatients	10 (7.6)	4 (40)
	Outpatient	123 (92.4)	9 (7.3)
7 - 12	Inpatients	61 (14.2)	14 (23)
	Outpatient	370 (85.8)	71 (19.1)
13 - 24	Inpatients	27 (17.4)	3 (11.1)
	Outpatient	128 (82.6)	19 (14.8)
25 - 36	Inpatients	13 (14.3)	1 (7.7)
	Outpatient	78 (85.7)	6 (7.7)
37 - 48	Inpatients	17 (22.7)	1 (5.8)
	Outpatient	58 (77.3)	7 (12)
> 48	Inpatients	9 (10.7)	2 (22.2)
	Outpatient	75 (89.3)	6 (8)
ND	Inpatients	6	1
	Outpatient	0	0
Total (%)		975	144 (14.7)

(ND=not defined)

5.2 Genotyping:

All 144 ELISA-positive samples were genotyped using oligonucleotides specific for the B and C regions of the NoV genome of genogroups I and II. Of these, the NoV of 140 samples were characterized as belonging to genogroup II; with four samples there was no amplification with any of the oligonucleotide pairs used.

5.3 Sequencing and phylogenetic analysis:

For sequencing, we selected some representative samples. The samples were then separated according to the date of collection into four quarters, and at least one sample from each quarter was chosen according to the intensity of the *amplicon* band observed on an agarose gel after RT-PCR. 16 samples from Uberaba and 17 samples from Uberlândia were chosen, totaling 33 samples representing all the years studied (Table 9).

Table 9: Distribution of the representative samples from Uberaba and Uberlândia that were sequenced.

Year Quarters	2005	2006	2007	2008	2009	2010	2011
1st Quarter	-	UB47	X	UB387	UB460 IP138	IP292 IP302 IP340	X
2nd Quarter		UB115 UB117	UB297	UB406	IP151	IP335 IP336	IP473
3rd Quarter	UB12	UB178	UB301 UB349	UB420 IP28	UB488 IP169	IP400	IP504 IP516
4th Quarter	UB14	UB254	UB357	IP101	IP249 IP244	X	IP514

(X) represents quarters in which no positive sample was detected, and (-) period not analyzed.

From the sequencing reactions, 33 consensus sequences were generated. The sequences contained 336 base pairs corresponding to the ORF1-ORF2 junction region (RpRd-VP1). The sequences were submitted to the *BLAST* tool (http://blast.ncbi.nlm.nih.gov/), and compared to other sequences by similarity. Analysis of the sequences generated confirmed the RT-PCR results; all the sequenced samples were found to belong to genogroup II, with 31 (94%) being characterized as genotype GII.4, one sample (3%) as genotype GII.3, and one sample (3%) as genogroup GII.6. The partial sequence of ORF2, which encodes the VP1 protein, was used to assemble the phylogenetic tree. The consensus sequences were then aligned by Clustal W using the Geneious 6.0.5 ® program and the similarity between the NoVs of those selected was analyzed. Some samples proved to be identical and in this case only one of them was used to build the phylogenetic tree (table 10); 17 sequences were thus aligned with the reference sequences obtained from *GenBank*.

Table 10: Comparison between the similarity of the sequenced samples for the choice of samples for phylogenetic analysis. Samples from Uberaba are represented by the letters UB, and samples from Uberlândia are represented by the letters IP.

Samples selected for phylogeny	Samples with identical sequences
	UB460
IP151	IP336
	UB301
IP400	UB349
	UB357
	UB47
IP514	UB117
UB387	IP269
	IP292
	IP244
	IP302
UB14	IP504
	IP516
	UB12
	UB406
	UB420

An analysis of the phylogenetic tree showed that 94% of the sequences analyzed in this study were representatives of NoV belonging to the GII.4 subgenotype, the largest number of which were similar to the 2006b/Nijmegen prototype, followed by the 2004/Hunter subgenotype (figure 14). Two sequences analyzed in this study (6%) belong to NoV GII.6 (3%) and NoV GII.3 (3%), but these strains are not related to other subgenotypes and have not been linked to epidemic outbreaks (figure 14).

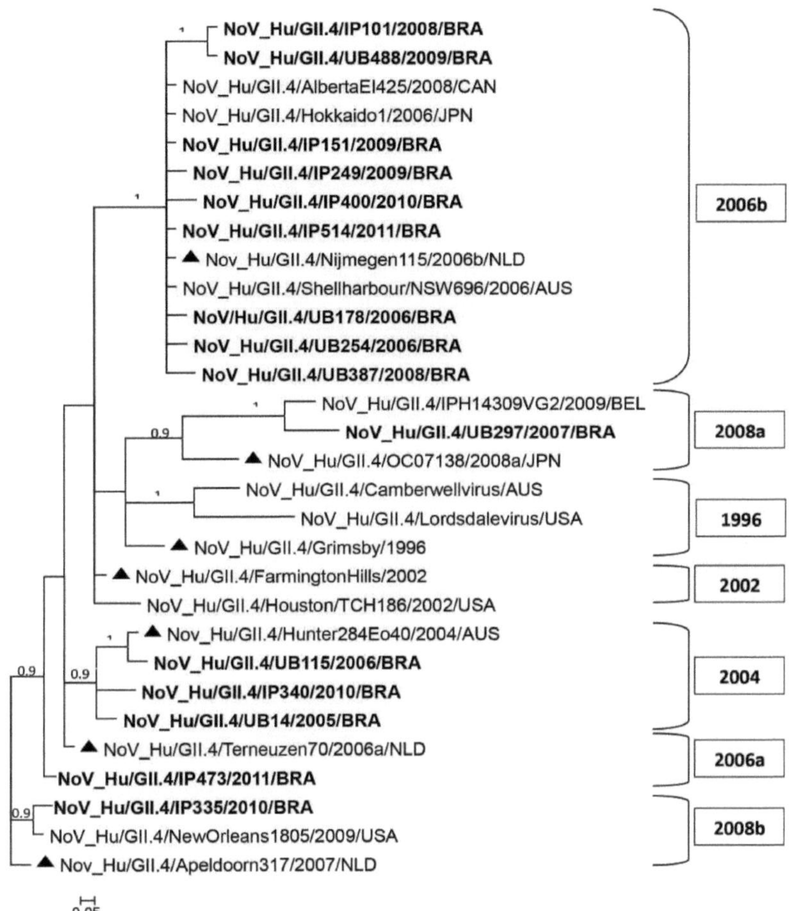

Figura 14: Phylogenetic tree of norovirus GII.4, built with a sequence of 280nt of the C region belonging to the ORF-2 encoding the VP1 protein, assembled with the *MrBayes 3.1.2*® program, by Bayesian methodology with 1,500,000 generations, *Tamura-2-parameter* model defined by *Jmodel test 2.1.1*® . The markings (A) represent the reference sequences of the GII.4 subgenotypes.

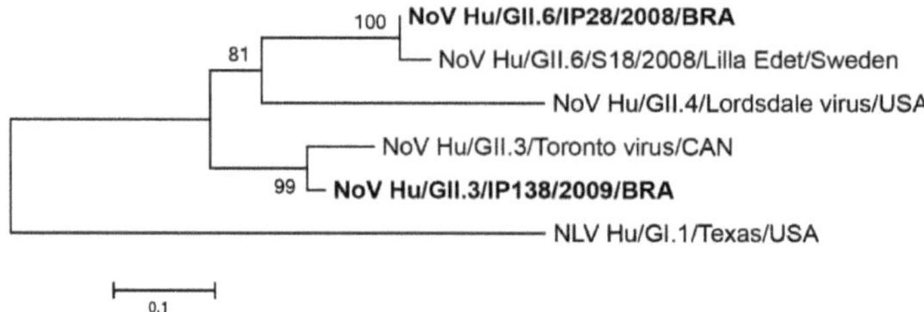

Figura 15: Phylogenetic tree of norovirus GII.3 and GII.6, built with a sequence of 280nt of the C region belonging to the ORF-2 encoding the VP1 protein, Maximum-likelihood methodology using the *Tamura-2-parameter* model defined by *Jmodel test 2.1.1*® with a *bootstrap* value of 2000 replicates.

6 DISCUSSION

Viruses are the most important agents of endemic and epidemic gastroenteritis worldwide, with NoV and RV being the main representatives (GLASS *et al.*, 2001). NoVs were the first viral agents associated with gastrointestinal disease to be discovered; however, they have long been considered secondary agents of gastroenteritis, after RVs, which have always had the highest prevalence rates (MORILLO *et al.*, 2011). With the introduction of the RV vaccine (G1P[8] Rotarix™) into the childhood vaccination schedule (RIBEIRO *et al.*, 2008), the prevalence of RV infections fell in the first year after introduction (LANZIERI *et al.*, 2011), which may have given NoV a selective advantage, putting this virus in the spotlight, especially as a cause of gastroenteritis in children under five years of age. However, the prevalence of these viruses in various regions of Brazil and the world is still not well established, underestimating their epidemiology in the population (BORGES *et al.*, 2006). This is the first epidemiological study on NoVs carried out in the Triângulo Mineiro region, where the prevalence of NoV infection was investigated in the two largest cities in this region.

The overall percentage of positivity in the study was 14.7%, 16.8% in Uberaba and 13.0% in Uberlândia. These prevalence rates are in line with those reported in various studies in Brazil and around the world. Epidemiological studies carried out in various Brazilian regions published in recent years have shown prevalence rates with a wide range varying from 7.6% (ANDREASI *et al.*, 2008) to 39.7% (RIBEIRO *et al.*, 2008) found in the states of Mato Grosso do Sul and Espirito Santo, respectively. Other studies conducted in Brazil have shown similar rates: BORGES *et al.* (2006) detected a percentage of 8.6% in the Brazilian Midwest; MORILLO *et al.* (2008) obtained a rate of 15.7% in the state of Sao Paulo; GEORGIADIS *et al.*, (2010) found 13.2% of positive samples in Porto Alegre-RS, and ARAGÂO *et al* (2010) found 12.5% in samples from children in Belém- PA. In other countries, prevalence rates have also varied widely, with 6.4% in Vietnam (NGUYEN *et al.*, 2008),   9.9% in Pakistan (PHAN *et al.*, 2004), 8.2% in the Republic of China.

(CHEN *et al.*, 2007), 12.0% in Nicaragua (BUCARDO *et al.*, 2008), 13.2% in Japan (NAKANISHI *et al.*, 2009), 15.0% in Korea (YOON *et al.*, 2008) and 17.0% in Mexico (KO *et al.*, 2005). In European countries, the average prevalence of NoV among children affected by gastroenteritis is 11.0%, with rates ranging from 6.0% to 20.0%, as reported in England and Finland, respectively (LOPMAN et al., 2002). These figures should be analyzed with caution. Although the prevalence rates in various localities in Brazil and in several countries around the world are very close, these rates could be even higher, due to the fact that many cases have not yet been reported (GEORGIADIS *et al.*, 2010). The percentage of norovirus positivity was higher than the percentage of rotavirus positivity (12.0%) previously reported for these samples (DULGHEROFF *et al.*, 2012

). It is worth noting that all the samples analyzed in this study came from cases of childhood gastroenteritis whose doctors requested diagnosis only for rotavirus, which has always been considered the main etiological agent of childhood gastroenteritis. Diagnosis of norovirus infections is not commonly requested due to the limited availability of reagents to detect this agent and also because historically it has not been a virus closely associated with infectious childhood gastroenteritis, but rather with gastroenteritis that affects individuals of all ages. However, in recent years the prevalence of norovirus infection in the etiology of childhood gastroenteritis has been increasing, perhaps due to the greater availability of diagnostic methodologies and an increase in the number of studies carried out. This fact shows the importance of epidemiological studies, which can be used as sources of data and information in public health, and can influence medical management and the decision to request laboratory diagnosis for other agents, so that the etiology of gastroenteritis cases can be elucidated more completely.

The highest prevalence rates found in this study occurred in Uberlândia in 2010 (20.0%) and in Uberaba in 2008 (19.2%) (table 2). A decrease in the number of samples received between 2006 and 2009 was also observed in Uberaba (figure 8); and the same decrease can be seen in Uberlândia between 2009 and 2011. However, despite the gradual drop in the number of samples received, there was an increase in the percentage of positive samples observed in this city in 2008, 2009 and 2010 (table 2), showing that the prevalence of this agent may be increasing in this region, and further prospective studies are needed in order to better characterize the prevalence of NoV.

Graphs 1 and 2 show that the samples did not follow a seasonal pattern; in 2006 and 2007 in Uberaba, there was an increase in the number of positive samples between the months of May and October, which are the driest and coldest months of the year in the Triângulo Mineiro region; the same could not be observed for 2008 and 2009, when there was a regular distribution throughout all the months of these years. In Uberlândia, no relationship was observed between positivity and the temperatures and rainfall indices measured in this region in any period (figure 9).

When looking at the percentage of positive samples overlapping all the years in the two locations (figures 10 and 11), it can be seen that Uberaba had two peaks of NoV infection, one in January, which had high rainfall levels as well as higher temperatures, and the other in August and September, with lower temperatures and the lowest rainfall levels observed during the year. In Uberlândia, the period with the highest prevalence was the driest and hottest months of the year, which comprised February, March and April.

BORGES *et al.* (2006) found NoV throughout the year in the central region of Brazil, but with greater prevalence in the months of September to March, a period of high relative humidity. This region has climatic characteristics similar to those found in the Triângulo Mineiro. In Porto Alegre-

RS, GEORGIADIS *et al* (2010) observed higher viral detection rates in spring, with 29.7% of positivity found in November. SOARES *et al* (2007) found no seasonality pattern in a study carried out between 1998 and 2005 in Rio de Janeiro. ANDREASI *et al* (2008) also found NoV circulating throughout the year in Campo Grande-MS, but with a slight increase in prevalence in March and April. In other countries, such as Japan and Kenya, a seasonal pattern was also not found (KOGAWA *et al.*, 1996; NAKATA et *al.,* 1998). In France, LORROT *et al* (2010) observed two NoV peaks during the two years of the study and both peaks occurred during the coldest months of the year, with the first peak in January and the second peak in September. Similar results were found in another study of Pakistani children in 2004, where two peaks of infection were found that coincided with the coldest months of the year, one in March and the other between September and October (PHAN *et al.*, 2004). In a review including twelve studies carried out in Europe, NoV were predominantly found during the colder months of the year, with an increase in prevalence in the months of October and November, with a peak in January and decreasing prevalence between the months of May and June (LOPMAN, BROWN, KOOPMANS, 2002). LEE *et al* (2008) investigated the presence of NoV in two seasons in South Korea, summer and winter, and found a detection rate of 21.7% and 17.3%, respectively.

The lack of a seasonal pattern can be explained, in the case of Brazil, by the fact that the winter is not so harsh, which is not the case in other countries, such as the USA and countries on the European continent, where the winter has a significant drop in temperature. In these regions, the intense cold causes the population to be confined to small spaces, leading to outbreaks; in Brazil, temperatures vary very little throughout the year, especially in the triângulo mineiro region, which causes the population to frequent more airy spaces, which does not lead to outbreaks related to cold periods (GEORGIADIS *et al.*, 2010).

NoV infections occurred in all age groups (figure 13), but the most affected was between 7 and 12 months, which was also observed by PHAN *et al* (2004), ANDREASI et *al* (2008), ARAGÀO *et al* (2010) and LORROT *et al* (2010). These results together show that children in the first year of life are more affected by both diarrhea and NoV infection. In this study, the highest positivity rates occurred in the first year of life, with 62.6% and 73.9% of infections occurring in Uberaba and Uberlândia respectively, which has also been demonstrated by other studies (PANG *et al.*, 1999; ROCKX *et al.*, 2002);

HANSMAN *et al.*, 2004 ; PHAN et *al.*, 2004 ; RIBEIRO *et al.,* 2008 ; NAKANISHI et *al.*, 2009 ; FERREIRA *et al.,* 2010 ; TRAN *et al., 2010*). However, KAPLAN *et al.* (1982) and BORGES *et al.* (2006) did not find a pattern of positivity linked to a specific age group, but rather a percentage of positivity with a similar distribution at all ages; and NGUYEN *et al.* (2008) in Japan observed a higher prevalence of NoV infections in children in the second year of life.

In our sample, considering Uberaba and Uberlândia, patients treated in outpatient clinics represented the majority of those affected by diarrhea (85.3%). However, when considering only those infected with norovirus, inpatients had a slightly higher prevalence (18.1%) compared to outpatients (14.7%). This can also be seen by analyzing the two locations separately (tables 7 and 8). SOARES et al. (2007) and VICTORIA et al. (2007) reported similar rates in Rio de Janeiro, with 15.4% and 20.0%, respectively. FERREIRA et al. (2010), in a four-year study carried out in the state of Rio de Janeiro, found a percentage of 30.2% and observed a percentage drop over the years studied among outpatients affected by NoV. On the other hand, they observed a gradual increase in the percentage among hospitalized patients. RIBEIRO et al. (2008) found a percentage of 39.7% among hospitalized patients in Espirito Santo, and also reported that NoVs were the most important cause of diarrhea among hospitalized children in that state during the study period, when compared to RV, with rates ranging from 5.4% to 30.0%.

In other countries, the highest percentages of positivity were also among hospitalized children. In France, LORROT et al. (2010) observed a percentage of 12.2%, corroborating the results published by TRAN et al. (2010), also carried out in France, in which 13.0% of cases were reported among hospitalized children. CUNLIFFE et al. (2010), in England, and BUCARDO et al. (2008), in Nicaragua, published results close to those found in France, with 16.0% and 15.0%, respectively. In three studies carried out in Thailand, NoV positivity rates among hospitalized children ranged from 8.1% to 14.1% (GUNTAPONG et al., 2004; HANSMAN et al., 2004; KHAMRIN et al., 2007). Higher rates were observed by TRANG et al., (2012) in Vietnam (36.0%), and by COLOMBA et al. (2007) in Italy, with 48.4%. These data suggest that norovirus has a higher prevalence of infection among hospitalized children than among those treated in outpatient clinics, in the same way as rotavirus (DULGHEROFF et al., 2012), which confirms the importance of this agent as a cause of more serious gastroenteritis, leading to a significant number of hospitalizations and, consequently, hospital costs.

Considering the relationship between the age of the patients, NoV positivity and the origin of the samples, it was observed that in the first two years of life there was a higher prevalence of norovirus positivity among hospitalized children (21.4%), when compared to those treated in outpatient clinics (15.9%), with a greater percentage difference observed among children aged between zero and six months. LORROT et al. (2010) also observed similar data in a study carried out in France, where 86.8% of hospitalized NoV-positive children were under two years old and 39.4% were aged between zero and six months. Similar data was observed by TRANG et al. (2012) in Vietnam. In Nicaragua, BUCARDO et al. (2008) found 16.2% positivity among hospitalized children under two years of age, but the highest percentages were observed among children between seven and 12 months. This study, as well as others mentioned here, showed that the highest rates of diarrhea and NoV positivity were observed among children between seven and 12 months of age (figures 12 and 13), which suggests that this infection occurs shortly after weaning, or when anti-norovirus antibodies are no longer present at protective levels. On the other hand, it should be noted that children between the ages of zero and six months, even in the presence of these antibodies, show a more severe manifestation of the disease, often requiring hospitalization.

Through analysis of the sequences obtained by sequencing, the NoV of the samples from Uberaba were characterized as being of the GII.4 genotype, while in Uberlândia most of the NoV of the samples were also classified as GII.4, but two samples were classified as GII.3 and GII.6 (figure 10B). Thus, the GII.4 genotype was the most prevalent in the entire study (94.0%), a fact also observed in other studies carried out in Brazil (CASTILHO, *et al.*, 2006 ; VICTORIA et al., 2007 ; CAMPOS *et al.*, 2008 ; FERREIRA *et al.,* 2008 ; BARREIRA et al., 2010) and in several other countries (LOPMAN et al., 2004 ; BULL *et al.,* 2006; ALLEN et al., 2008; SIEBENGA *et al.,* 2009; PANG *et al.,* 2010). In Brazil, NoVs belonging to GII.3 were also found in the states of Espirito Santo and Rio de Janeiro, while viruses of the GII.6 genotype were found in the states of Pernambuco, Espirito Santo, Minas Gerais, Rio de Janeiro and Sao Paulo (FIORETTI *et al.*, 2011).

The high prevalence of GII.4 worldwide has been explained as being the result of point mutations occurring among circulating viruses, giving rise to new variant forms, replacing existing strains and thus allowing them to persist in the population (LINDESMITH *et al.*, 2008). This genotype was the first to be recognized as an epidemic strain in the mid-1990s in the United States (NOEL et al., 1999). In the following years several epidemic outbreaks were reported, and the responsible strains were genetically characterized as being subgenotypes GII.4. In 1995-1996 outbreaks occurred in the USA and the Netherlands (VINJÉ et al., 1997) and later the strain was named US95/96 (NOEL et al., 1999). During the period from 2000 to 2004 there were outbreaks in Europe and the USA, where the Farmington Hill strain was responsible for 80% of gastroenteritis outbreaks, replacing the US95/96 strain (WIDDOWSON et al., 2004). In 2004, a new variant associated with outbreaks in Australia, the Netherlands, Japan and Taiwan called "Hunter" was detected (BULL et al., 2005). Between 2004 and 2006, and between 2007 and 2008, several strains responsible for outbreaks across Europe were reported in the FBVE (Foodborne viruses in Europe network) database, including the 2006a strain found in England, 2006b causing outbreaks in Spain (KRONEMAN et al., 2006 ; PANG et al., 2010). All the strains causing outbreaks in Europe and worldwide are genetically linked and have been internationally classified as variants of the GII.4 genotype: 1996 (US96), 2002 (Farmington Hill), 2004 (Hunter), 2006a , 2006b, 2008a and 2008b (PANG et al.,

2010).

Through partial sequencing of the C region of the main capsid protein (VP1), it was possible to observe through phylogenetic analysis that the NoV from the samples collected in the Triângulo Mineiro region were associated with five variants of the GII.4 genotype: 2004, 2006a, 2006b, 2008a, and 2008b, cited by PANG et al (2010), with the majority of these strains belonging to the 2006b subgenotype. (Figure 14). In Brazil, a review study found NoV GII.4 in nine of the 13 Brazilian states studied, which is the main genotype associated with gastroenteritis circulating in this country. In this study, the following variants were detected: 2004 (Hunter), 2006a and 2006b (FIORETTI et al., 2011). These variants were also identified in the present study, suggesting that these genetic variants of this genotype circulate in different regions of the country; better and more complete monitoring and a better and broader knowledge of the molecular characterization and genetic diversity of the NoV circulating in our country are needed.

7   CONCLUSIONS

- This study showed the circulation of NoV in the Triângulo Mineiro region

- NoV was more prevalent than RV, when compared to data obtained in a study for this viral agent in this region.

- NoV infection showed no seasonality during the period studied.

- There was a higher prevalence among children aged seven to 12 months and a higher percentage among hospitalized children under two years of age.

- Partial sequencing of the C region of the VP1 protein showed that the GII.4 genotype was predominantly circulating.

- The majority of NoV circulating in Triangulo Mineiro belonged to the Nijimegem subgenotype (2006b) followed by the Hunter subgenotype (2004).

- This epidemiological and molecular study of NoVs was the first to be carried out in the triangulo mineiro region, and sequential studies should be continued in order to gain a better understanding of the behavior of these viruses in the community, health institutions, nurseries and schools, mainly for the creation of epidemiological control measures aimed at prophylaxis and reducing transmission.

8 BIBLIOGRAPHICAL REFERENCES

ADLER JL, ZICKL R. **Winter vomiting disease.**J Infect Dis 1969; 119: 668-73.

AGUS sg, DOLIN r, WYATT RG, TOUSIMIS AJ, NORTHRUP RS 1973. **Acute infectious nonbacterial gastroenteritis: intestinal histopathology.** Ann Inter Med 79: 18-25.

ALAM, N. H., ASHRAF, H. **Treatment of infectious diarrhea in children.** Paediatr Drugs, v.5, n.3, p.151-65. 2003

ALLEN, D.J., J.J. GRAY, C.I. GALLIMORE, J. XERRY and M. ITURRIZA-GOMRA. 2008. **Analysis of amino acid variation in the P2 domain of the GII.4 norovirus VP1 protein reveals putative variant-specific epitopes.** PLoS ONE. 3(1);e1485.

ANDO T, MONROE SS, GENTSCH JR, JIN Q, LEWIS DC, GLASS RI 1995. **Detection and differentiation of antigenically distinct small round-structured viruses** (Norwalk-like viruses). J Clin Microbiol 38: 64-71.

ANDO, T.; NOEL, J. S.; FRANKHAUSER, R. L. **Genetic classification of "Norwalklike Viruses".** The Journal of Infectious Diseases, Chicago, v. 181, Suppl. 2, p. 336348, 2000.

ANDREASI M,S,A. CARDOSO D,D. FERNANDES S,M. TOZETTI I,A. BORGES A,M,T. FIACCADORI F,S. ET Al. **Adenovirus, calicivirus and astrovirus detection in fecal samples of hospitalizedchildren with acute gastroenteritis from Campo Grande,** MS, Brazil. Mem Inst Oswaldo Cruz 2008;103(7):741-4.

ARAGÂO, G C.; OLIVEIRA, D. S.; SANTOS, M. C.; MASCARENHAS, J. D. P.; OLIVEIRA, C. S.; LINHARES, A. C.; GABBAY, Y. B. **Molecular characterization of norovirus, sapovirus and astrovirus in children with acute gastroenteritis from Belém, Parà, Brazil.** Rev Pan-Amaz Saude, v. 1, p. 149-158, 2010.

ATMAR, R.L., OPEKUN, A.R., GILGER, M.A., ESTES, M.K., CRAWFORD, S.E., NEILL, F.H., GRAHAM, D.Y., 2008. **Norwalk virus shedding after experimental human infection.** Emerg Infect Dis 14, 1553-1557. PMC2609865

ATMAR, R. L.; ESTES, M. K. **Diagnosis of noncultivatable gastroenteritis viruses, the human caliciviruses.** Clinical Microbiology Reviews, Washington, v. 14, n. 1, p. 15-37, 2001.

ATMAR RL, ESTES MK 2001. **Diagnosis of noncultivatable gastroenteritis viruses, the human caliciviruses.** Clin Microbiol Rev 14 (1): 15-37.

BARKER J, VIPOND IB, BLOOMFIELD SF. **Effects of cleaning and disinfection in reducing the spread of Norovirus contamination via environmental surfaces.** J Hosp Infect. 2004; 58(1):

42-9.

BARREIRA, D. M. P. G.; FERREIRA, M. S. R.; FUMIAN, T. M.; CHECON, R.; SADOVSKY, A. D. I.; LEITE, J. P. G.; MIAGOSTOVICH, M. P.; SPANO, L. C. **Viral load and genotypes of noroviruses in symptomatic and asymptomatic children in Southeastern Brazil**. J Clin Virol, v. 47, p. 60-64, 2010.

BELLIOT G, SOSNOVTSEV SV, MITRA T, HAMMER C, GARFIELD M, GREEN KY. **In vitro proteolytic processing of the MD145 norovirus ORF1 nonstructural polyprotein yields stable precursors and products similar to those detected in calicivirus-infected cells**. J Virol. 2003; 77(20): 10957-74.

BERKE, T., B. GOLDING, X. JIANG, D. W. CUBITT, M. WOLFAARDT, A. W. SMITH, and D. O. MATSON. 1997. **Phylogenetic analysis of the Caliciviruses**. J Med Virol 52:419-24.

BERTOLOTTI-CIARLET A, CRAWFORD SE, HUTSON AM, ESTES MK. **The 3' end of Norwalk virus mRNA contains determinants that regulate the expression and stability of the viral capsid protein VP1: a novel function for the VP2 protein**. J Virol. 2003; 77(21): 11603-15.

BERTOLOTTI-CIARLET A, WHITE LJ, CHEN R, PRASAD BVV, ESTES MK 2002. **Structural requirements for the assembly of norwalk virus-like particles**. J Virol 76 (8): 4044-4055.

BORGES AM, TEIXEIRA JM, COSTA PS, GIUGLIANO LG, FIACCADORI FS, FRANCO RDE C, BRITO WM, LEITE JPG, CARDOSO DD. **Detection of calicivirus from fecal samples from children with acute gastroenteritis in the West Central region of Brazil**. Men Inst Oswaldo Cruz 2006; 101: 721-724

BOSCHI-PINTO C, VELEBIT L, SHIBUYA K. **Estimating child mortality due to diarrhoea in developing countries**. Bull World Health Organ. 2008; 86:710-717.

BUCARDO F, NORDGREN J, CARLSSON B, PANIAGUA M, LINDGREN PE, ET AL. **Pediatric norovirus diarrhea in Nicaragua**. J Clin Microbiol. 2008;46:2573- 2580

BULL, R. A., HANSMAN, G S., CLANCY, L. E., TANAKA, M. M., RAWLINSON, W. D. & WHITE, P. A. (2005). **Norovirus recombination in ORF1/ORF2 overlap**. Emerg Infect Dis 11, 1079-1085

BULL R. A., TU ET, MCIVER CJ, RAWLINSON WD, WHITE PA. **Emergence of a new norovirus genotype II.4 variant associated with global outbreaks of gastroenteritis**. J Clin Microbiol 2006; 44: 327-33.

CAMPOS, G. S.; MOREAU, V. H.; BANDEIRA, A.; BARBERINO, G.; ALMEIDA, P. F.; AMADOR, D. M.; DE LIMA, M. O.; SARDI, S. I. **Molecular detection and genetic diversity of** norovirus in hospitalized young adults with acute gastroenteritis in Bahia, Brazil. Arch. Virol., v. 153, n. 6, p. 1125-1129, apr. 2008.

CAO, S., Z. LOU, M. TAN, Y. CHEN, Y. LIU, Z. ZHANG, X.C. ZHANG, X. JIANG, X. LI, AND Z. RAO. 2007. **Structural basis for the recognition of blood group trisaccharides by norovirus**. J. Virol. 81(11):5949-57.

CARDOSO DDP, BORGES AMT. **Human calicivirus**. Rev Pat Trop 2005; 34: 1726.

CASTILHO, J. C.; MUNFORD, V.; RESQUE, H. R.; FAGUNDES-NETO, U.; VINJÉ, J.; RACS, M. L. **Genetic diversity of No among children with gastroenteritis in Sao Paulo state, Brazil**. J Clin Microbiol, v. 44, p. 3947-3953, 2006.

CAUL EO, APPELENTON H. **The electron microscopical and physical characteristics of small round human fecal viruses: an interim scheme for classification**. J Med Virol 1982; 9: 257-265.

CAUL EO 1996. **Viral gastroenteritis: small round structured viruses, caliciviruses and astroviruses**. Part I. The clinical and diagnostic perspective. J Clin Pathol 49: 87480.

CAUL EO. **Viral gastroenteritis: small round structured viruses, caliciviruses and astroviruses**. Part II. The epidemiological perspective. J Clin Pathol. 1996; 49(12): 95964.

CHEN, R., NEILL, J. D., et al. **X-ray structure of a native calicivirus: structural insights into antigenic diversity and host specificity**. Proc Natl Acad Sci U S A, v.103, n.21, May 23, p.8048-53. 2006.

CHEN SY, CHANG YC, LEE YS, CHAO HC, TSAO KC, LIN TY, KO TY, TSAI CN, CHIU CH. **Molecular epidemiology and clinical manifestations of viral gastroenteritis in hospitalized pediatric patients in Northern Taiwan**. J Clin Microbiol 2007; 45:2054-7.

CHENG AC, MCDONALD JR, THIELMAN NM. **Infectious diarrhea in developed and developing countries** J Clin Gastroenterol 2005; 39:757-73.

CHIBA S, NAKATA S, NUMATA-KINOSHIRA K, HONMA S 2000. **Sapporo virus: history and recent findings.** J Infect Dis 181: S303-8.

CILLI, AUDREY et al. **Molecular characterization of rotavirus and norovirus strains: a 6-year study (2004-2009).** J. Pediatr. (Rio J.). 2011, vol.87, n.5, pp. 445-449.

CLARK B, MCKENDRICK M. **A review of viral gastroenteritis** Curr Opin Infect Dis 2004 17:461-9.

COLOMBA, C., DE GRAZIA, S., et al. **Viral gastroenteritis in children hospitalized in Sicily, Italy**. Eur J Clin Microbiol Infect Dis, v.25, n.9, Sep, p.570-5. 2006.

CUBITT, W. D., MITCHELL, D. K., et al. **Application of electronmicroscopy, enzyme immunoassay, and RT-PCR to monitor an outbreak of astrovirus type 1 in a paediatric bone marrow transplant unit**. J Med Virol, v.57, n.3, Mar, p.313-21. 1999.

CUBITT WD, MCSWIGGAN DA, MOORE W. **Winter vomiting disease caused by calicivirus.** J Clin Pathol 32: 786-793, 1979.

CUNLIFFE. N. A, BOOTH. J.A, ELLIOT. C, LOWE. S. J, SOPWITH. W, KITCHIN. N, NAKAGOMI. O, NAKAGOMI. T, HART. C. A, M**. Healthcare-associated Viral Gastroenteritis among Children in a Large Pediatric Hospital, United Kingdom** Emerg Infect Dis. 2010 January; 16(1): 55-62. PMCID: PMC2874353 doi: 10.3201/eid1601.090401.

DANIELS NA, BERGMIRE-SWEAT DA, SCHWAB KJ. **A foodborne outbreak of gastroenteritis associated with Norwalk-like viruses: first molecular trace back to deli sandwiches contaminated during preparation**. J Infect Dis 2000; 181:14671470.

DEDMAN D, LAURICHESSE H, CAUL EO, WALL PG. **Surveillance of small round structured virus (SRSV) infection in England and Wales, 1990-5**. Epidemiol Infect. 1998; 121(1): 139-49.

DOLIN, R., LEVY, A. G., et al. **Viral gastroenteritis induced by the Hawaii agent. Jejunal histopathology and serologic response**. Am J Med, v.59, n.6, Dec, p.761-8. 1975.

DOLIN R, BLACKLOW NR, DUPONT H. **Biological properties of Norwalk agent of acute infectious nonbacterial gastroenteritis**. Proc Soc Exp Biol Med 1972; 140: 578583.

DOLIN R, BLACKLOW NR, DUPONT H. **Transmission of acute infectious nonbacterial gastroenteritis to volunteers by oral administration of stool filtrates**. J Infect Dis1971; 123:307-312.

DONALDSON, E.F., LINDESMITH, L.C., LOBUE, A.D., BARIC, R.S., 2008. **Norovirus pathogenesis: mechanisms of persistence and immune evasion in human populations**. Immunological Reviews, 225, 190-211.

DONALDSON E. F., LINDESMITH L, LOBUE A, BARIC R (2010) **Viral shapeshifting: norovirus evasion of the human immune system**. Nature Reviews Microbiology 8: 231-241. doi: 10.1038⁄nrmicro2296.

DRUMMOND AJ, ASHTON B, BUXTON S, CHEUNG M, COOPER A, DURAN C, ET AL. **Geneious v6.0.6.** Biomatters Ltd, Auckland, New Zealand. 2011.

DULGHEROFF AC, FIGUEIREDO EF, MOREIRA LP, MOREIRA KC, MOURA LM, GOUVÊA VS,

DOMINGUES AL. **Distribution of rotavirus genotypes after vaccine introduction in the Triângulo Mineiro region of Brazil: 4-Year follow-up study.** *Journal of Clinical Virology* (July 2012), doi:10.1016/j.jcv.2012.06.003.

ESTES MK, BALL JM, GUERRERO RA, OPEKUN AR, GILGER MA, PACHECO SS. **Norwalk virus vaccines: Challenges and progress.** J Infect Dis. 2000; 181(Suppl 2): S367-S373.

FERNANDEZ JM, GÓMEZ JB. **Norovirus infections.** Enferm Infecc Microbiol Clin. 2010 Jan;28 Suppl 1:51-5. doi: 10.1016/S0213-005X(10)70009-4 . Review. Spanish. PMID: 20172424.

FERREIRA, M. S. R.; VICTORIA, M.; CARVALHO-COSTA, F. A.; VIEIRA, C. B.; XAVIER, M. P. T P.; FIORETTI, J. M.; ANDRADE, J.; VOLOTÂO, E. M.; ROCHA, M.; LEITE, J. P. G.; MIAGOSTOVICH, M. P. **Surveillance of norovirus infections in the state of rio de janeiro, brazil 2005-2008.** J. Of med. Virol., v. 82, p. 1442-1448, 2010.

FERREIRA, M. S. R.; XAVIER, M. P. T. P.; FUMIAN, T. M.; VICTORIA, M.; OLIVEIRA, S. A.; PENA, L. H. A.; LEITE, J. P. G.; MIAGOSTOVICH, M. P. **Acute gastroenteritis cases associated with noroviruses infection in the state of rio de janeiro.** J. Of Med. Virol., v. 80, n. 2, p. 338-344, feb. 2008.

FIORETTI, J. M.; FERREIRA, M. S. R.; VICTORIA, M.; VIEIRA, C. B.; XAVIER, M. P. T. P.; LEITE, J. P. G.; MIAGOSTOVICH, M. P. **Genetic diversity of norovirus in brazil.** Mem. Inst. Oswaldo Cruz, v. 106, n. 8, p. 942-947, 2011.

FRANKHAUSER, R. L.; MONROE, S. S.; NOEL, J. S.; HUMPHREY, C. D.; BRESSE, J. S.; PARASHAR, U. D.; ANDO, T.; GLASS, R. I. **Epidemiologic and molecular trends of "Norwalk-like Viruses" associated with outbreaks of gastroenteritis in the United States.** The Journal of Infectious Diseases, Chicago, v. 186, n. 1, p. 1-7, 2002.

FRANKHAUSER R. L, NOEL JS, MONROE SS, ANDO T, GLASS RI 1998. **Molecular epidemiology of norwalk-like viruses in outbreaks of gastroenteritis in the United States.** J Infect Dis 178: 1571- 8.

GABBAY YB, GLASS RI, MONROE SS, CARCAMO C, ESTES MK, MASCARENHAS JD, LINHARES AC 1994. **Prevalence of antibodies to norwalk virus among amerindian isolated amazonian communities.** Am J Epidemiol 139: 738-33.

GABBAY YB, GLASS RI, MONROE SS. **Prevalence of antibodies to Norwalk virus among Amerindians in isolated Amazonian communities.** Am J Epidemiol 1994; 139: 728-733.

GALLIMORE CI, BARREIROS MAB, BROWN DWG, NASCIMENTO JP, LEITE JPG 2004. **Norovirus associated with acute gastroenteritis in a children's day care facility in Rio de**

Janeiro, Brazil. Bras J Med Biol Research 37: 321-326.

GEORGIADIS, S.; PILGER, D. A.; PEREIRA, F.; CANTARELLI, V V **Molecular evaluation of norovirus in patients with acute gastroenteritis.** Revista da sociedade brasileira de medicina tropical,v. 43, n. 3, p. 277-280, mai-jun. 2010.

GLASS, R. I., BRESEE, J., et al. **Gastroenteritis viruses: an overview.** Novartis found symp, v.238, p.5-19; discussion 19-25. 2001.

GLASS, R. I., NOEL, J., et al. **The epidemiology of enteric caliciviruses from humans: a reassessment using new diagnostics.** J Infect Dis, v.181 Suppl 2, May, p.S254-61. 2000.

GLASS PJ, WHITE LJ, BALL JM, LEPARC-GOFFART I, HARDY ME, ESTES MK. **Norwalk virus open reading frame 3 encodes a minor structural protein.** J Virol 2000; 74: 6581-91.

GORDON I, INGRAHAM HS, KORNS RF. **Transmission of epidemic gastroenteritis to human volunteers by oral administration of fecal isolates.** J Exp Med 1947; 86: 409-422.

GORDON, I. 1955. **The nonamebic nonbacillary diarrheal disorders.** Amer J Trop Med. 4:739-55.

GRAHAM, D.Y., X. JIANG, T. TANAKA, A.R. OPENKUN, H.P. MADORE AND M.K. ESTES. 1994. **Norwalk virus infection of volunteers: new insights based on improved assays.** J Infect Dis. 170(1):34-43.

GREEN K. Y., ANDO T, BALAYAN MS, BERKE T, CLARKE IN, ESTES MK, MATSON DO, NAKATA S, NEIL JD, STUDDERT MJ, THIEL H-J 2000. **Taxonomy of the caliciviruses.** J Infect Dis 181: S322-30.

GREEN, K. Y. **Caliciviridae: the noroviruses.** In: knipe d. M; howley, p. M. (eds.). Fields in virology.5 ed. Philadelphia: lippincott-raven publishers,2007. Vol 1, p. 949979.

GREEN K. Y., CHANOCK RM, KAPIKIAN AZ. HUMAN CALICIVIRUSES. IN: KNIPE DM, HOWLEY PM, CHANOCK RM, MELNICK JL, MONATH TP, ROIZMAN B, STRAUS SE. (EDS.) **Fields in Virology Philadelphia,** Lippincott- Raven Publishers, 4th ed. 2001 vol 1, p. 875-93.

GREEN J, GALLIMORE CI, NORCOTT JP, LEWIS D, BROWN DWG. **Broadly reactive reverse transcriptase polymerase chain reaction for the diagnosis of SRSV** - associated gastroenteritis. J Med Virol 47: 392-398, 1995.

GREENBERG HB, KAPIKIAN AZ 1978A. **Detection of norwalk agent antibody and antigen by solid-phase radioimmunoassay and immune adherence hemagglutination assay.** JAVMA 173: 620-23.

GREENBERG HB, WYATT RG, VALDESUSO J, KALICA AR, LONDON WT, CHANOCK RM, KAPIKIAN AZ 1978B. **Solid-phase microtiter radioimmunoassay for detection of the norwalk strain of acute nonbacterial, epidemic gastroenteritis virus and its antibodies.** J Med Virol 2: 97-108.

GUNTAPONG R, HANSMAN GS, OKA T, OGAWA S, KAGEYAMA T, PONGSUWANNA Y, KATAYAMA K. **Norovirus and sapovirus infections in Thailand.** Jpn J Infect Dis. 2004; 57(6): 276 - 278. [PubMed: 15623956].

HALE AD, TANAKA TN, KITAMOTO N. **Identification of an epitope common to genogroup I "Norwalk-like viruses".** J Clin Microbiol 2000; 38:1656-1660.

HANSMAN GS, DOAN LT, KGUYEN TA, OKITSU S, KATAYAMA K, OGAWA S, ET AL. **Detection of norovirus and sapovirus infection among children with gastroenteritis in Ho Chi Minh City, Vietnam.** Arch Virol. 2004; 149(9): 1673-88.

HANSMAN, GS, KATAYAMA, K, MANEEKARN, N, PEERAKOME, S, KHAMRIN, P, TONUSIN, S, OKITSU, S, NISHIO, O, TAKEDA, N, USHIJIMA, H (2004) **Genetic diversity of norovirus and sapovirus in hospitalized infants with sporadic cases of acute gastroenteritis in Chiang Mai, Thailand.** J Clin Microbiol 42: pp. 1305-1307.

HARDY ME. **Norovirus protein structure and function.** FEMS Microbiol Lett 2005 1; 253:1-8.

HARDY, M.E. AND M.K. ESTES, 1996. **Completion of the Norwalk virus genome sequence.** Virus Genes, 12: 287-29

HUELSENBECK JP, RONQUIST F. MRBAYES: **Bayesian inference of phylogenetic trees.** Bioinformatics. 2001 Aug;17(8):754-5.

ICTV2005 . **Index of Viruses.** Available at: www.ncbi.nih.gov/ICTVdb/Ictv/fs_calic.htm. Accessed on: February 2013.

INOUYE, S.; YAMASHITA, K.; YAMADERA, S.; YOSHIKAWA, M.; KATO, N.; OKABE, N. **Surveillance of viral gastroenteritis in Japan: pediatric cases and outbreak incidents.** The Journal of Infectious Diseases, Chicago, v. 181, Suppl. 2, p. 270-274, 2000.

INSTITUTO ADOLFO LUTZ E CENTRO DE VIGILÂNCIA EPIDEMIOLOGICA "PROFESSOR ALEXANDRE VRANJAC". **Diarrhea and rotavirus.** Rev. Saùde Pùblica. 2004, vol.38, n.6 pp. 844-845.

INTERNATIONAL COMMITTEE ON TAXONOMY OF VIRUSES - ICTV. ICTVDB: **The universal virus database: version 4.** Available at:

<http://www.ncbi.nlm.nih.gov/ICTVdb/ICTVdB/>. Accessed on: February 2013.

JIANG X, ESPUL C, ZHONG WM, CUELLO H, MATSON DO 1999. **Characterization of a novel human calicivirus that may be a naturally occurring recombinant**. Arch Virol 144: 2377-2387.

JIANG X, GRAHAM DY, WANG K, ESTES MK. **Norwalk virus genome cloning and characterization**. Science 1990; 250:1580-1583.

JIANG X, WANG M, WANG K, ESTES MK 1993. **Sequence and genomic organization of norwalk virus**. Virology 195: 51-61.

JIANG X, WANG J, GRAHAM DY, ESTES MK. **Detection of Norwalk virus in stool by polymerase chain reaction**. Clin Microbiol 1992a; 30: 2529-2534.

JIANG X, WANG M, GRAHAM DY, ESTES MK. **Expression, self-assembly, and antigenicity of Norwalk virus capsid protein**. J Virol 1992b; 66: 6527-6532.

JIANG X, WILTON N, ZHONG WM, FARKAS T, HUANG PW, BARRET E, GUERRERO M, RUIZ-PALACIOS, GREEN KY, GREEN J, HALE AD, ESTES MK, PICKERING LK, MATSON DO 2000. **Diagnosis of human caliciviruses by use of enzyme immunoassays**. J Infect Dis 181: S349-59.

JORDAN WS, GORDON I. **A study of illness in a group of Cleveland families. VII.Transmission of acute nonbacterial gastroenteritis to volunteers: Evidence for two different etiological agents**. J Exp Med 1953; 98: 461-475.

KAPIKIAN AZ. Norwalk **and Norwalk-like viruses**. In: Kapikian AZ, ed. Virus Infections of the Gastrointestinal Tract. New York: Marcel Dekker, 1994: 471-518.

KAPIKIAN AZ. **"Overview of viral gastroenteritis"** (1996). Arch. Virol. Suppl. 12: 7-19.

KAPIKIAN AZ. **The discovery of the 27-nm Norwalk virus: an historic perspective**. J Infect Dis. 2000; 181(Suppl 2): S295-302.

KAPIKIAN, A. Z. **Viral infections of the gastrointestinal** tract. 2 ed. New York: Marcel-Deker, 1994.

KAPIKIAN AZ, ESTES MK, CHANOCK RM. **Norwalk group of viruses**. In: Fields Knipe BN, Howley PM et al., eds. Virology. 1996; 3rd Vol 2. New York: Lippincott-Raven, 1996: 783-810.

KAPIKIAN AZ, WYATT RG, DOLIN R, THORNHILL TS, KALICA ARL, CHANOCK RM. **Visualization by immune electron microscopy of 27 nm particle associated with acute infections nonbacterial gastroenteritis**. J Virol 1972; 10: 10751081.

KAPLAN JE, GARY GW, BARON RC, SINGH N, SCHONBERGER LB, FELDMAN R, GREENBERG HB 1982A. **Epidemiology of Norwalk gastroenteritis and the role of Norwalk virus in outbreaks of acute gastroenteritis**. Ann Intern Med 96: 756-761.

KAPLAN JE, FELDMAN R, CAMPBELL D, LOOKABAUGH C, GARY W 1982B. **The frequency of a norwalk-like pattern of illness in outbreaks of acute gastroenteritis**. Am J Public Health 72: 1329-1332.

KARST SM. **Pathogenesis of noroviruses, emerging RNA viruses**. Viruses. 2010;2:748-781.

KATAYAMA K, HANSMAN GS, OKA T, OGAWA S, TAKEDA N **Investigation of norovirus replication in a human cell line**. Arch Virol 2006; 151:1291-308. doi: 10.1007/s00705-005-0720-9.

KATAYAMA, K., H. SHIRATO-HORIKOSHI, S. KOJIMA, T. KAGEYAMA, T. OKA, F. HOSHINO, S. FUKUSHI, M. SHINOHARA, K. UCHIDA, Y. SUZUKI, T. GOJOBORI, AND N. TAKEDA. 2002. **Phylogenetic analysis of the complete genome of 18 Norwalk-like viruses**. Virology 299(2):225-239.

KESWICK, B. H.; SATTERWHITE, T. K.; JOHNSON, P. C. **Inactivation of Norwalk virus in drinking water by chlorine**. Appl. Environ. Microbiol., v. 50, p. 261-264, 1985.

KHAMRIN P, MANEEKARN N, PEERAKOME S, et al. **Genetic diversity of noroviruses and sapoviruses in children hospitalized with acute gastroenteritis in Chiang Mai, Thailand**. J Med Virol 2007; 79: 1921-6.

KHAMRIN P, OKAME M, THONGPRACHUM A, NANTACHIT N, NISHIMURA S, OKITSU S, ET AL. (2011) **A single-tube multiplex PCR for rapid detection in feces of 10 viruses causing diarrhea**. J Virol Methods 173(2):3-390. PubMed PMID: 21349292.

KITTIGUL L, POMBUBPA K, TAWEEKATE Y, DIRAPHAT P, SUJIRARAT D, KHAMRIN P, USHIJIMA H 2010. **Norovirus GII-4 2006b variant circulating in patients with acute gastroenteritis in Thailand during a 2006-2007 study**. J Med Virol 82: 854-860.

KOBAYASHI S, SAKAE K, SUZUKI Y, SHINOZAKI K, OKADA M, ISHIKO H, KAMATA K, SUZUKI K, NATORI K, MIYAMURA T, TAKEDA N. **Molecular cloning, expression, and antigenicity of Seto virus belonging to genogroup I Norwalk-like viruses**. J Clin Microbiol. 2000; 38(9): 3492-4.

KOGAWA K, NAKATA S, UKAE S, ADACHI N, NUMATA K, MATSON DO, ESTES MK, CHIBA S. **Dot blot hybridization with a cDNA probe derived from the human calicivirus Sapporo 1982 strain**. Arch Virol. 1996; 141(10): 1949 - 1959.

KO G, GARCIA C, JIANG ZD, et al. **Noroviruses as a cause of traveler's diarrhea among students from the United States visiting Mexico.** J Clin Microbiol. 2005;43:6126-9.

KOJIMA S, KAGEYAMA T, FUKUSHI S, HOSHINO FB, SHINOHARA M, UCHIDA K, NATORI K, TAKEDA N, KATAYAMA K. **Genogroup-specific PCR primers for detection of Norwalklike viruses.** J Virol Methods 2002; 100: 107-114.

KOOPMANS, M.; DUIZER, E. **Foodborne viruses: an emerging problem.** Int. J. Food Microbiol., v. 90, n. 1, p. 23-41, 2004.

KOOPMANS M. **Molecular Epidemiology of human enteric caliciviruses in The Netherlands.** In: Gastroenteritis Viruses. Chichester: Wiley. 2001; 197-218.

KOOPMANS, M., VON BONSDORFF, C. H., ET AL. **Foodborne viruses.** FEMS Microbiol Rev, v.26, n.2, Jun, p.187-205. 2002.

KOU, X., Q. WU, J. ZHANG; AND H. FAN. 2006. **Rapid detection of noroviruses in fecal samples and shellfish by nucleic acid sequence-based amplification.** J. Microbiol. 44:403-408.

LANZIERI TM, LINHARES AC, COSTA I, KOLHE DA, CUNHA MH, ORTEGA- BARRIA E, COLINDRES RE 2011. **Impact of rotavirus vaccination on childhood deaths from diarrhoea in Brazil.** int J infect Dis 15: e206-210.

LEE, H.; PARK, Y.; KiM, M.; JEE, Y.; cHEON, D.; JEONG, H. S.; KO, G. **Development of a Latex Agglutination Test for Norovirus Detection.** J. Food Microbiol., v. 48, n. 4, p. 419-425, 2010.

LE PENDU J (2004) **Histo-blood group antigen and human milk oligosaccharides: genetic polymorphism and risk of infectious diseases.** Adv Exp Med Biol 554: 135143.

LE PENDU, J., N. RUVOEN-CLOUET, E. KINDBERG; AND L. SVENSSON. 2006. **Mendelian resistance to human norovirus infections.** Semin. immunol. 18:375-386.

LiNDESMiTH Lc, DONALDSON EF, LOBUE AD, cANNON JL, ZHENG DP, ViNJE J, BARic RS. **Mechanisms of GII.4 norovirus persistence in human populations.** PLoS Med 2008 Feb; 5(2): e31.

LINDESMITH L, MOE C, MARIONNEAU S, RUVOEN N, JIANG X, LINDBLAD L, STEWART P, LEPENDU J, BARIC R. **Human susceptibility and resistance to Norwalk virus infection.** Nat Med 2003 May; 9(5): 548-53.

LINHARES AC, STUPKA JA, CIAPPONI A, BARDACH AE, GLUJOVSKI D, ARUJ PK, ET AL. **Burden and typing of rotavirus group A in Latin America and the Caribbean: systematic** review and meta-analysis. Rev Med Virol. 2011;21(2):89- 109.

LOPMAN BA, BROWN DW, KOOPMANS M 2002. **Human caliciviruses in Europe**. J Clin Microbiol 24: 137-160.

LOPMAN BA, REACHER MH, VIPOND IH, SARANGI J, BROWN DW **Clinical manifestation of norovirus gastroenteritis in health care settings**. Clin Infect Dis 2004; 39: 318-24.

LORROT M, BON F, EL HAJJE MJ, AHO S, WOLFER M, GIRAUDON H, KAPLON J, MARC E, RAYMOND J, LEBON P, POTHIER P, GENDREL D. **Epidemiology and clinical features of gastroenteritis in hospitalized children: prospective survey during a 2-year period in a Parisian hospital, France**. Eur J Clin Microbiol Infect Dis. 2011 Mar;30(3):361-8. doi: 10.1007/s10096-010-1094-9.

MARIONNEAU, S. N. RUVOËN, B. LE MOULLAC-VAIDYE, M. CLEMENT, A. CAILLEAU-THOMAS, G. RUIZ-PALACOIS, P. HUANG, X. JIANG, AND J. LE PENDu. 2002. **Norwalk virus binds to histoblood group antigens present on gastroduodenal epithelial cells of secretor individuals**. Gastroenterology. 122(7);1967-77.

MARKS PJ, VIPOND IB, CARLISLE D, DEAKIN F, FEY RE, CAuL EO 2000. **Evidence for airborne transmission of norwalk-like virus (NVL) in a hotel restaurant**. Epidemiol Infect 124: 481-487.

MATSuNO S, SAWADA R, KIMuRA K, SuZuKI H, YAMANISHI S, SHINOZAKI K, SuGIEDA M & HASEGAWA A (1997). **Sequence analysis of SRSV in fecal specimens from an epidemic of infantile gastroenteritis, October to December 1995, Japan**. Journal of Medical Virology, 52: 377-380.

MINISTRY OF HEALTH. **Health Surveillance Secretariat**: Epidemiological Surveillance Guide. 7 ed. - Brasilia, 2009.

MINISTRY OF HEALTH. **Epidemiological situation of acute diarrhea in children**.

2012Access onlineat : http://portal.saude.gov.br/portal/saude/ profissional/area.cfm?id area=1549

MORENO-ESPINOSA S, FARKAS T, JIANG X. **Human caliciviruses and pediatric gastroenteritis.** Semin Pediatr Infect Dis 2004; 15: 237-245.

MORILLO, S. G.; CILLI, A.; CARMONA, R. C. C.; TIMENETSKY, M. C. S. T. **Identification and molecular characterization of norovirus in Sao Paulo State, Brazil.** Brazilian J. of Microbiol., v. 39, n. 4, p. 619-622, 2008.

MORILLO, S. G.; LUCHS, A.; CILLI, A.; RIBEIRO, C. D.; CALUX, S. J.; CARMONA, R. C. C.; TIMENETSKY, M. C. S. T. **Large gastroenteritis outbreak due to norovirus GII in Sao Paulo, Brazil, summer 2010.** Rev. Inst. Med. Trop. Sao Paulo, v. 53, n. 2, p. 119-120, 2011.

MORILLO, S. G.; TIMENETSKY, M. C. S. T. **Norovirus: an overview.** Rev. Assoc. Med. Bras., v. 57, n. 4, p. 453-458, 2011.

MURATA, T., N. KATSUSHIMA, K. MIZUTA, Y. MURAKI, S. HONGO, AND Y. MATSUZAKI. 2007. **Prolonged norovirus shedding in infants <or=6 months of age with gastroenteritis.** Pediatri. Infect. Dis. J. 26(1);46-9.

NAKANISHI K, TSUGAWA T, HONMA S, NAKATA S, TATSUMI M, YOTO Y, TSUTSUMI H 2009. **Detection of enteric viruses in rectal swabs from children with acute gastroenteritis attending the pediatric outpatient clinics in Sapporo, Japan.** J Clin Virol 46: 94-97.

NAKATA S, HONMA S, NUMATA K, KOGAWA K, UKAE S, ADACHI N, JIANG X, ESTES MK, GATHERU Z, TUKEI PM, CHIBA S 1998. **Prevalence of human calicivirus infections in Kenya as determined by enzyme immunoassays for three genogroups of the virus.** J Clin Microbiol 36(11): 3160- 3163.

NGUYEN TA, HOANG L. PHAM LE D, HOANG KT, OKITSU S, MIZUGUCHI M, et al. Norovirus and sapovirus infections among children with acute gastroenteritis in Ho Chi Minh City during 2005-2006. **J Trop Pediatr. 2008;54:102-13.**

NICOLLIER-JANNOT B, OGIER A, PIROTH L, POTHIER L, KOHLI E. **Recombinant virus-like particles of a norovirus (genogroup II strain) administered intranasally and orally with mucosal adjuvants LT and LT (R192G) in BALB/c mice induce specific humoral and cellular Th/Th2-like immune responses.** Vaccine 2004: 22 (9-10):1079-86.

NOEL, J.S., ANDO, T., LEITE, J.P., GREEN, K.Y., DINGLE, K.E., ESTES, M.K., SETO, Y., MONROE, S.S., GLASS, R.I., 1997. **Correlation of patient immune responses with genetically characterized Small Round-Structured Viruses involved in outbreaks of nonbacterial acutegastroenteritis in the United States, 1990-1995.** Journal of Medical Virology, 53, 372-383.

NOEL JS, FRANKHAUSER RL, ANDO T, MONROE SS, GLASS RI 1999. **Identification of a distinct common strain of "norwalk-like viruses" having a global distribution.** *J Infect Dis* 179: 1334-44.

NUMATA K, HARDY ME, NAKATA S, CHIBA S, ESTES MK. **Molecular characterization of morphologically typical human calicivirus Sapporo.** Arch Virol. 1997; 142(8): 1537 - 1552.

OKHUYSEN, P.C., X. JIANG, L. YE, P.C. JOHNSON, AND M.K. ESTES. 1995. **Viral shedding** and fecal IgA response after Norwalk virus infection. J. Infect. Dis. 171(3):566-9.

ORLANDI PP, SILVA T, MAGALHAES GF, ALVES F, CUNHA RPA, DURLACHER RR, PEREIRA DA SILVA LH 2001. **Enteropathogens associated with diar-rheal disease in infants of poor urban areas of Porto Velho, Rondônia: a preliminary study**. Mem Inst Oswaldo Cruz 96: 621-625.

PANG X, JOENSUU J, VESIKARI T. **Human calicivirus-associated sporadic gastroenteritis in Finnish children less than two years of age followed prospectively during a rotavirus vaccine trial**. Pediatr Infect Dis J 18: 420-426, 1999.

PANG XL, HONMA S, NAKATA S, VESIKARI T. **Human caliciviruses in acute gastroenteritis of young children in the community**. J Infect Dis. 2000; 181 Suppl 2: S288-94.

PANG XL, PREIKSAITIS JK, WONG S, LI V, LEE BE 2010. **Influence of novel norovirus GII.4 variants on gastroenteritis outbreak dynamics in Alberta and the northern territories, Canada between 2000 and 2008**. *PLoS ONE 5*: e11599

PARASHAR UD, DOW L, FRANKHAUSER RL, HUMPREY CD, MILLER J, ANDO T, WILLIAMS KS, EDDY CR, NOEL JS, INGRAM T, BRESEE JS, MONROE SS, GLASS RI 1998. **An outbreak of viral gastroenteritis associated with consumption of sandwich: implications for the control transmission by food handlers**. Epidemiol Infect 121: 615-621.

PARASHAR U, D., MONROE SS. **"Norwalk-like viruses" as a cause of foodborne disease outbreaks**. Rev Med Virol. 2001; 11(4): 243-52.

PARKS CG, MOE CL, RHODES D, LIMA A, BARRET L, TSENG F, BARIC R, TALAL A, GUERRANT R 1999. **Genomic diversity of norwalk like viruses (NLVs): pediatric infections in a brazilian shantytown**. J Med Virol 58: 426-34.

PATEL MM, WIDDOWSON MA, GLASS RI, AKAZAWA K, VINJE J AND PARASHAR UD. **Systematic literature review of role of noroviruses in sporadic gastroenteritis**. Emerg Infect Dis 2008 Aug; 14(8): 1224-31.

PHAN TG, OKAME M, NGUYEN TA, MANEEKARN N, NISHIO O, OKITSU S, USHIJIMA H. **Human astrovirus, norovirus (GI, GII), and sapovirus infections in Pakistani children with diarrhea**. J Med Virol. 2004; 73(2): 256-61.

POSADA D. jModelTest: **Phylogenetic Model Averaging**. Mol. Biol. Evol. 25(7):1253-1256. 2008.

PLATT AR, WOODHALL RW, GEORGE AL JR. **Improved DNA sequencing quality and efficiency using an optimized fast cycle sequencing protocol**. Biotechniques. 2007 Jul;43(1):58,

60, 62.

PRASAD B, V, HARDY ME, DOKLAND T, BELLA J, ROSSMANN MG, ESTES MK. **X-ray crystallographic structure of the Norwalk virus capsid**. Science 1999; 286: 287-90.

PRASAD, B. V., M. E. HARDY, X. JIANG, AND M. K. ESTES. 1996. **Structure of Norwalk virus.** Arch. Virol. Suppl. 12:237-242.

RADFORD, A.D. ET AL. 2007. **Feline calicivirus.** Veterinary Research. 38: 319-335.

REIMANN HA, HODGES JH, PRICE AH. **Epidemic diarrhea, nausea, and vomiting of unknown cause.** JAMA 1945a; 127: 1-6.

REIMANN HA, HODGES JH, PRICE AH. **The cause of epidemic diarrhea, nausea, and vomiting (viral dysentery?).** Proc Soc Exp Biol Med. 1945b; 59: 8-9.

RIBEIRO, L, R., et al. **Hospitalization due to norovirus and genotypes of rotavirus in pediatric patients, state of Espirito Santo**. Mem. Inst. Oswaldo Cruz [online]. 2008, vol.103, n.2 [cited 2013-05-13], pp. 201-206 .

RICHARDS, A. F., B. LOPMAN, A. GUNN, A. CURRY, D. ELLIS, H. COTTERILL, S. RATCLIFFE, M. JENKINS, H. APPLETON, C. I. GALLIMORE, J. J. GRAY, AND D. W. G. BROWN. 2003. **Evaluation of a commercial ELISA for detecting Norwalklike virus antigen in faeces**. J. Clin. Virol. 26:109-115.

RICHARDS GP, WATSON MA, FANKHAUSER RL, MONROE SS. **Genogroup I and II noroviruses detected in stool samples by real-time reverse transcription- PCR using highly degenerate universal primers**. Appl Environ Microbiol 2004; 70: 7179-84.

ROCKX, B. H.; DE WIT, M.; VENNEMA, H.; VINJÉ, J.; DE BRUIN, E.; VAN DUYNHOVEN, Y.; KOOPMANS, M. **Natural history of human calicivirus infection: a prospective cohort study.** Clinical Infectious Diseases. v. 35, n. 3, p. 24653, Aug 2002.

SDIRI-LOULIZI K, AMBERT-BALAY K, GHARBI-KHELIFI H, SAKLY N, HASSINE M, ET AL. (2009) **Molecular epidemiology of norovirus gastroenteritis investigated using samples collected from children in Tunisia during a four-year period: detection of the norovirus variant GGII.4 Hunter as early as January 2003**. J Clin Microbiol 47: 421-429. doi: 10.1128/JCM.01852-08

SIEBENGA, J. J.; VENNEMA, H.; ZHENG, D.; VINJÉ, J.; LEE, B. E.; PANG, X.; HO, E. C. M.; LIM, W.; CHOUDEKAR, A.; BROOR, S.; HALPERIN, T.; RASOOL, N. B. G.; HEWITT, J.; GREENING, G. E.; JIN, M.; DUAN, Z.LUCERO, Y.; O'RYAN, M.; HOEHNE, M.; SCHREIER, E.; RATCLIFF, R. M.; WHITE, P. A.; IRITANI, N.; REUTER, G.; KOOPMANS, M. **Norovirus**

Illness Is a Global Problem: Emergence and Spread of Norovirus GII.4 Variants, 2001-2007. The J. of Infect. Dis., v. 200, p. 802-812, 2009.

SOARES, C. C.; SANTOS, N.; BEARD, R. S.; ALBUQUERQUE, M. C.; MARANHAO, A. G; ROCHA, L. N.; RAMiREZ, M. L.; MONROE, S. S.; GLASS, R. I.; GENTSCH, J. **Norovirus Detection and Genotyping for Children with Gastroenteritis, Brazil.** Emerg. Infect. Dis., v. 13, n. 8, p. 1244-1246, 2007.

STRAUB TM, HONER ZU, BENTRUP K, OROSZ-COGHLAN P, DOHNALKOVA A, MAYER BK, BARTHOLOMEW RA, et al. **In vitro cell culture infectivity assay for human noroviruses.** Emerg Infect Dis 2007; 13: 396-403.

TACKET, C.O., M.B. SZTEIN, G.A. LOSONSKY, S.S. WASSERMAN, AND M.K. ESTES. 2003. **Humoral, mucosal, and cellular immune responses to oral Norwalk virus like particles in volunteers.** Clin Immunol. 108(3):241-7.

TACKET CO. **Plant-derived vaccines agains diarrheal diseases.** Vaccine 2005; 23 (15):1866-9.

TALAL AH, MOE CL, LIMA AA, WEIGLE KA, BARRETT L, BANGDIWALA SI, ESTES MK, GUERRANT RL. **Seroprevalence and seroincidence of Norwalk-like virus infection among Brazilian infants and children.** J Med Virol 2000; 61: 117-124.

TAN, M., AND JIANG, X. **Norovirus gastroenteritis, increased understanding and future antiviral options.** Curr Opin Investig Drugs 2008 Feb; 9(2): 146-51.

TAN, M., AND. JIANG, X. **Norovirus-host interaction: implications for disease control and prevention.** Expert Rev. Mol. Med. 9:1-22. 2007.

TAN, M. AND. JIANG, X. **The p domain of norovirus capsid protein forms a subviral particle that binds to histo-blood group antigen receptors.** J. Virol. 79(22);14017-30. 2005.

TAN M, HEGDE RS, JIANG X. **The P domain of norovirus capsid protein forms dimer and binds to histo-blood group antigen receptors.** J Virol 2004; 78: 6233-42.

TAN, M., HUANG P, MELLER J, ZHONG W, FARKAS T, JIANG X. **Mutations within the P2 domain of norovirus capsid affect binding to human histo-blood group antigens: evidence for a binding pocket.** J Virol. 2003; 77(23): 12562-71.

TAN, M.; XIA, M.; CHEN, Y.; BU, W.; HEGDE, R.S.; MELLER, J.; LI, X.; JIANG, X. **Conservation of Carbohydrate Binding Interfaces - Evidence of Human HBGA Selection in Norovirus Evolution.** PLoS ONE 2009.

THORNTON AC, JENNINGS-CONKLIN KS, MCCORMICK MI. **Noroviruses: agents in outbreaks of acute gastroenteritis.** Disaster Management Response 2004; 2: 4-9.

TIMENETSKY MC, KISIELIUS JJ, GRISI SJ, ESCOBAR AM, UEDA M, TANAKA H 1993. **Rotavirus, adenovirus, astrovirus, calicivirus and small round virus particles in feces of children with and without acute diarrhea, from 1987 to 1988, in the greater Sao Paulo.** Rev Inst Med Trop Sao Paulo 35: 275-80.

TOPOROVISKI, M. S., CHIEFFI, P. P. ET AL. **Acute diarrhea in children under 3 years of age: recovery of enteropathogens in fecal samples from patients compared to the control group.** Jornal de Pediatria, v.72, n.2, p.97-104. 1999.

TRAN A, TALMUD D, LEJEUNE B, JOVENIN N, RENOIS F, PAYAN C, LEVEQUE N, ANDREOLETTI L 2010. **Prevalence of rotavirus, adenovirus, norovirus, and astrovirus infections and coinfections among hospitalized children in northern France.** J Clin Microbiol 48: 1943-1946.

TRANG NV, LUAN LE T, KIM-ANH LE T, HAU VT, NHUNG LE TH, PHASUK P, ET AL. **Detection and molecular characterization of noroviruses and sapoviruses in children admitted to hospital with acute gastroenteritis in Vietnam.** J Med Virol. 2012;84:290-7.

UNICEF (The United Nations Children's Fund). **Diarrhoea: why children are still dying and what can be done.** Geneva; 2009

VASICKOVA P, DVORSKA L, LORENCOVA A, PAVLIK I 2005. **Viruses as a cause of foodborne diseases:a review of the literature.** Vet Méd 50:89-104.

VICTORIA, M.; CARVALHO-COSTA, F. A.; HEINEMANN, M. B.; LEITE, J. P.; MIAGOSTOVICH, M. **Prevalence and molecular epidemiology of noroviruses in hospitalized children with acute gastroenteritis in Rio de Janeiro, Brazil,** 2004. Pediatr Infect Dis J, Rio de Janeiro, v. 26, p. 602-606, 2007

VICTORIA, M.; GUIMARAES, F.; FUMIAN, T.; FERREIRA, F.; VIEIRA, C.; LEITE, J. P. G.; MIAGOSTOVICH, M. P. **Evaluation of an adsorption-elution method for detection of astrovirus and norovirus in environmental waters.** J. of Virol. Methods, v. 156, n. 1-2, p. 73-76, 2009.

VIDAL, R.; SOLARI, V.; MAMANI, N.; JIANG, X.; VOLLAIRE, J.; ROESSLER, P.; PRADO, V.; MATSON, D. O.; O'RYAN, M. L. **Caliciviruses and foodborne gastroenteritis, Chile.** Emerg. Infect. Dis., v. 11, n. 7, p. 1134-1137, 2005.

VINJE J, ALTENA SA, KOOPMANS MPG 1997. **The incidence and genetic variability of small round-structured viruses in outbreaks of gastroenteritis in the Netherlands.** *J Infect Dis* 176: 1374-8.

VINJÉ J, KOOPMANS MP. **Molecular detection and epidemiology of small round- strutured viruses in outbreaks of gastroenteritis in the Netherlands.** J Infect Dis 1996; 174: 610-615.

XAVIER, M.P.T.P. et al. **Detection of caliciviruses associated with acute infantile gastroenteritis in Salvador, an urban center in Northeast Brazil.** Braz J Med Biol Res, May 2009, vol.42, no.5, p.438-444. ISSN 0100-879X

WIDDOWSON,M. A.; CRAMER, E. H.; HADLEY, L.; BRESEE, J. S.; BEARD, R. S.; BULENS, S. N.; CHARLES, M.; CHEGE, W.; ISAKBAEVA, E.; WRIGHT, J.G.; MINTZ, E.; FORNEY, D.; MASSEY, J.; GLASS, R. I.; MONROE, S. S. **Outbreaks of Acute Gastroenteritis on Cruise Ships and on Land: Identification of a Predominant Circulating Strain of Norovirus-United States, 2002.** J. of Infect. Diseases, v.190, p. 27-36, 2004.

WILHELMI, I., ROMAN, E., ET AL. **Viruses causing gastroenteritis.** Clin Microbiol Infect, v.9, n.4, Apr, p.247-62. 2003.

WORLD HEALTH ORGANIZATION - WHO. **Diarrhoeal Diseases.** Accessed July 2013 http://www.who.int/vaccine research/diseases/diarrhoeal/en/.

WYATT, R. G., R. DOLIN, N. R. BLACKLOW, H. L. DUPONT, R. F. BUSCHO, T. S. THORNHILL, A. Z. KAPIKIAN, AND R. M. CHANOCK. 1974. **Comparison of three agents of acute infectious nonbacterial gastroenteritis by cross-challenge in volunteers.** Journal of Infectious Diseases 129:709-14.

YAN H, YAGYU F, OKITSU S, NISHIO O, USHIJIMA H 2003. **Detection of norovirus (GI, GII), sapovirus and astrovirus in fecal samples using reverse transcription sigle-round multiplex PCR.** J Virol Methods 114: 37-44.

YOON JS, LEE SG, HONG SK, LEE SA, JHEONG WH, OH SS, OH MH, KO GP, LEE CH, PAIK SY. **Molecular epidemiology of norovirus infections in children with acute gastroenteritis in South Korea in November 2005 through November 2006.** J Clin Microbiol 2008; 46: 1474-7.

ZAHORSKY J. **Hyperemesis hemia or the winter vomiting disease.** Arch Pediatr 1929; 46: 391-395.

ZHENG DP, ANDO T, FANKHAUSER RL, BEARD RS, GLASS RI AND MONROE SS. **Norovirus classification and proposed strain nomenclature.** Virology 2006 Mar 15; 346(2): 312-23.

ZINTZ C, BOK K, PARADA E, BARNES-ELEY M, BERKE T, STAAT MA, AZIMI P, JIANG X, MATSON DO 2005. **Prevalence and genetic characterization of calicivirus among children**

hospitalized for acute gastroenteritis in the United States. Infect Gen Evolut 5: 281-90.

61

ANNEX 2: PREPARATION OF SOLUTIONS

Figure 1: Components of the *RIDASCREEN® Norovirus 3rd Generation* ELISA kit (R Biopharm AG Landwenrstr, Germany)

Kit components	Description
Plate	96 cavities coated with monoclonal antibodies against norovirus GI and GII
Diluent 1	Protein buffered NaCl solution; containing 0.1% NaN3 (ready for use)
Washing solution	10x phosphate buffered NaCl solution with 0.1% thimerosal (diluted 1:10 with distilled water)
Control +	Recombinant norovirus antigens
Conjugate 1	Biotin-conjugated antibodies against norovirus in protein solution stabilized, and 0.05% proclin 300.
Conjugate 2	Streptavidin-peroxidase conjugate in stabilized protein solution containing 0.05% proclin 300
Substrate	Hydrogen peroxide/TMB
Stop solution	Reaction blocker; 1N sulfuric acid.

item 2: Solutions used for nucleic acid extraction

2.1 Phosphate-buffered saline (PBS) (pH 7.4)

Sodium chloride	4g
Potassium chloride	0,1g
Sodium hydrogen phosphateanhydrous1	,2g
Dihydrogen phosphatepotassium phosphate	0,2g
Enough distilled water (q.s.p)	500mL

In a 1000mL beaker, the reagents were added, dissolved in 400mL of distilled water, homogenized with a magnetic stirrer and the pH adjusted to 7.4 with 36.5 to 38% (v/v) hydrochloric acid (HCl) PA before completing the final volume in a 500mL volumetric flask. The solution was transferred to a sealed flask, autoclaved at 121°C for 15 minutes and stored between 2 and 4 °C

2.2 Lid Tris-HCl 0.1 M (pH= 6.4)

Tris- hydroxymethyl-aminomethane1   .21g

Distilled water q.s.p   100mL

Tris and 80mL of distilled water were added to a 250mL beaker. The solution was homogenized with a magnetic stirrer and the pH adjusted to 6.4 with HCl PA before making up the final volume in a 100mL volumetric flask. The contents were transferred to a sealed flask, autoclaved at 121 °C for 15 minutes and stored between 2 and 4 °C

2.3 EDTA 0.2M (pH=8.0)

EDTA- ethylenediamine tetraacetic acid0   .744g

Distilled water q.s.   p10mL

EDTA and 8mL of distilled water were added to a 100mL beaker. The solution was stirred with a magnetic stirrer and the pH adjusted to 8.0 with 1N NaOH. The solution was transferred to a volumetric flask, adjusting the final volume with water and stored at 22-25°C in a bottle with a lid.

2.4 L6 cover

Guanidine isothiocyanate (GITC)   12g

EDTA 0.2M (pH=8.0)   2.2mL

Triton X-100   0,24mL

Tris-HCl 0.1M (pH=6.4) q.s.   p10mL

In a 100mL beaker, the reagents were added, diluted in 5mL of 0.1M Tris HCl (pH=6.4), homogenized with a magnetic stirrer, transferred to a volumetric flask and the final volume completed to 10mL. The solution was transferred to an amber bottle and stored between 22 and 25°C. The solution is stable for at least three weeks.

2.5 Lid L2

Guanidine isothiocyanate (GITC)   12g

Tris-HCl 0.1M (pH=6.4) q.s.   p10mL

The reagents were added to a 100mL beaker, diluted in 5mL of 0.1M Tris HCl (pH=6.4), homogenized with a magnetic stirrer, transferred to a volumetric flask and the final volume was made up to 10mL. The solution was transferred to an amber bottle and stored between 22 and 25°C. For up to three weeks.

2.6 Silica

Silicon dioxide (Sigma-Aldrich corp. St. Louis, Mo, USA®) 60g

Distilled water q.s.p   500mL

Water and 60g of silicon dioxide were added to a 500mL beaker. The solution was homogenized by pouring it into the beaker and leaving it to settle for 24 hours. 430mL of the supernatant was discarded and 500mL of distilled water was added to the silica. After sedimentation for 5 hours, 440mL of the supernatant was discarded. The pH was adjusted to 2.0 with 600μL of 37% HCl. The solution was aliquoted into 5mL tubes, autoclaved at 121°C for 15 minutes and stored protected from light at 22-25°C. For up to 6 months.

2.7 70% alcohol

Ethyl alcohol PA35   ,85mL

Distilled water   13,15mL

In a 100mL cylinder, 35.85mL of alcohol and 13.15mL of distilled water were mixed. The contents were homogenized by inversion, transferred to a stoppered bottle and stored between 2 and 4 °C.

2.8 Acetone PA

Ready for use

2.9   Tris-borate-EDTA (TBE) buffer 10x pH=8.4

Tris26   ,95g

Boric acid   13,75g

EDTA1   ,86g

Distilled water q.s.   p250mL

The reagents and 200mL of distilled water were added to a 500mL beaker. The solution was homogenized with a magnetic stirrer. The contents were transferred to a volumetric flask and the final volume was added and transferred to a sealed flask. The solution was stored between 2 and 4°C.

2.10 Tris-boro-EDTA lid 0.5x

The 10X TBE lid was diluted 1/20

2.11 Agarose gel 1.5%

Agarose   1,5g

| TBE buffer 0.5X pH=8.4 | 100mL |

Water and agarose were added to a vial with a lid. The mixture was then heated and homogenized and stored at room temperature.

2.12 Loading buffer

Glycerol	5mL
EDTA	0,04g
Bromophenol blue	0,025g
Sterile water q.s.p	5mL

Glycerol and EDTA were added to a 50mL *Falcon* tube and the pH was adjusted to 8.0. Bromophenol blue was then added. The solution was homogenized and stored between 2 and 8°C.

item 3: Preparation of the solutions used in the RT-PCR reactions

Table 1: Reverse transcription reaction solution

Concentration

Reagent	FinalVolume	(iiL)/reaction
5X reaction cap	1X	5µL
DTT 0.1M	0.	004M1µL
RNase OUT	40U	1µL
dNTP: dATP;dCTP;dGTP;dTTP 10mM	0.	4mM1µL
Random Hexamer (Invitrogen®) Reverse transcriptase	300ng1µL	
SuperScript II (Invitrogen®)	200U1µL	
Total	10µL	

3.1 Oligonucleotides

All the oligonucleotide pairs came from the factory freeze-dried and then diluted to a concentration of 200 µM. From this dilution, they were then diluted again with sterile water to 10 µM (the concentration used in the reactions), using 2.5 µL of the oligonucleotide with 47.5 µL of sterile water.

Primer pairs for genogroup 1		Primer pairs for genogroup 2	
FORWARD	**REVERSE**	**FORWARD**	**REVERSE**

1	MON 431	MON 433	COG2F	G2SKR
2	GISKF	GISKR	JV-12	ACAL-36
3	CAL-32	MO3-N	MON 381	MON 383
4	-	-	MON 432	MON 434

Table 2: PCR reaction and genotyping solution

| | | Concentration | |
Reagent		Final	Volume (uL)/reaction
Pure water			30.1μL
Reaction cap without MgCl2 10x		1X	5μL
MgCl2 50 μM		1.5 μM	1.5μL
dNTP: dATP;dCTP;dGTP;dTTP 10 μM		0.4 μM	2μL
Forward Oligonucleotide Pair 10 μM		0.2 μM	1μL
Reverse Oligonucleotide Pair 10 μM		0.2 μM	1μL
Taq DNA Polymerase 5U/μL (Invitrogen®)		0,4U	0.4μL
Total			40μL

Table 3: Sequencing reaction solution.

Reagent	Concentration	Volume / reaction
BigDye sequencing cap	5X	2 μL
Solution with *sense* or *antisense* primers	3.3μM	1 μL
BigDye *mix*	2.5X	0.3 μL
Purified DNA samples	-	*
Pure water	-	Make up to 10μL
Total	-	10 μL

*The amount of DNA that was added to the reaction was 3 to 10ng.

Printed by Books on Demand GmbH, Norderstedt / Germany